AF430909

A Textbook of

Research Methodologies and Biostatistics for Pharmacy Students

A Textbook of
Research Methodologies and Biostatistics for Pharmacy Students

Prof. K. P. R. Chowdary

PharmaMed Press
An imprint of Pharma Book Syndicate
A unit of BSP Books Pvt., Ltd.
4-4-309/316, Giriraj Lane,
Sultan Bazar, Hyderabad - 500 095.

A Textbook of Research Methodologies and Biostatistics for Pharmacy Students
by **Prof. K. P. R. Chowdary**

Published by

PharmaMed Press

An imprint of Pharma Book Syndicate

A unit of BSP Books Pvt. Ltd.

4-4-309/316, Giriraj Lane, Sultan Bazar, Hyderabad - 500 095.

Phone: 040-23445688, 23445600; Fax: 91+40-23445611

E-mail: info@pharmamedpress.com

www.pharmamedpress.com/pharmamedpress.net

ISBN: 978-93-89354-56-0

Dedicated to

My Teacher, Guide and Well-Wisher

Prof. V. Subba Rao

(1920-1990)

Preface

Biostatistics is a relatively new but rapidly expanding field of science, which find wide applications in pharmacy and pharmaceutical research. Pharmacy is a research based discipline. A knowledge of basic concepts of research, research methodologies, experimental designs and protocols and analysis of data resulting in good and meaningful interpretation are vital requirements for successful research in pharmacy.

The present textbook is the outcome of 40 years of teaching Biostatics and Research Methodologies to pharmacy students at Andhra University and other pharmacy colleges by the author. In addition to teaching, the author has applied the knowledge of statistical methods and research methodologies in organizing the pharmacy research work at PhD level and successfully guided 95 PhDs in pharmaceutical sciences at Andhra University, Acharya Nagarjuna University, Jawaharalal Nehru Technological University Hyderabad and Kakinada. Students of pharmacy find no textbook on biostatistics and research methodologies exclusively with pharmacy orientation and applications and as such the students are facing problems at all levels, U.G, P.G and PhD in better understanding and using statistical methods in their research and professional activities. This prompted the author to come out with this textbook.

This book has been designed to make the subject more interesting, more comprehensive and easy to grasp. The subject is presented in a modulated and graded manner, beginning with basic concepts and then gradually from the simple to advanced topics, making the students progress smooth, easy and pleasant. The uniqueness of the textbook is to include a number of solved problems and case studies at the end of each topic. Special emphasis is given on topics like experimental designs and protocols for human and animal studies, design of experiments (DOE), tests of significance including non - parametric tests, analysis of variance (ANOVA), optimization techniques, factorial experiments and optimization by factorial designs, correlation and regression, probit analysis and determination of LD_{50} and ED_{50}. A chapter on patentable research in pharmacy, patenting procedures with examples is also included.

The author is deeply grateful to the fellow teachers at Andhra University College of Pharmaceutical Sciences, Visakhapatnam, Vikas Institute of Pharmaceutical Sciences, Rajahmundry, KVSR Siddhartha College of Pharmaceutical Sciences, Vijayawada and School of Pharmaceutical Sciences, JNTUK, Kakinada for their valuable suggestions and help in the preparation of this textbook. The author specially acknowledge with deep sense of gratitude the support received from Dr. T. V. Narayana, Director and Principal Vikas Institute of Pharmaceutical Sciences, Rajahmundry and President, Indian Pharmaceutical Association, Mumbai. The author invites constructive comments and suggestions from readers to further improve this book in future revisions.

Prof. K. P. R. Chowdary

Contents

CHAPTER 3

Research Methodologies - II
(Designs for Preclinical and Clinical Studies)

Basic Concepts of Biostatistics

CHAPTER 7

Analysis of Variance

CHAPTER 8

Research Methodologies – III
(Designs for Laboratory Investigations)

CHAPTER 9

Research Methodologies - IV
(Factorial Experiments)

CHAPTER 10

Research Methodologies - V
(Doe and Optimization)

Non-Parametric Tests of Significance

Chapter 1

Introduction to Biostatistics

Biostatistics

Biostatistics is that branch of statistics which deals with problems in biological sciences. In other words, application of statistical methods to biological sciences is said to be biostatistics.

Biostatistics broadly deals with statistical applications in the context of biological problems including medicine, pharmacy and public health. Biostatistics is developed during the period of Sir Francis Galton (1822-1910) who is known as father of biometry. Measurement of biological aspects is known as biometrics and hence the terms biostatistics and biometry are synonymously used. Sir Francis Galton while examining the relationship between height of fathers and their sons based on heredity concept proposed concept of regression. Karl Pearson, basically a statiscian tried to apply statistical methods to biological problems. Later, R. A. Fischer was also attracted towards the application of statistics in biology and made significant contribution to the development of biostatistics.

To know Biostatistics, its significance and applications in pharmacy.

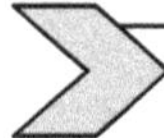 ## Applications of Biostatistics in Pharmacy

1. Biostatistics has been applied to a wide variety of problems in pharmacy, medicine and other health sciences.

2. Government organizations, research institutes and industry have been extensively using statistics and biostatistics. The census wing of the government extensively uses biostatistics in the analysis of data. In the interpretation of population characteristics such as birth rates, death rates, population growth rates, infant mortality rate, gross reproduction rate, age specific fertility, life expectancy etc. biostatistics is very much helpful. The ministry of health and family welfare also makes use of biostatistics while formulating policies related to medical care for child and women development, family planning, health and sanitation programs. The ministry of agriculture also depends much on biostatistics in planning for production of food grains and other commercial crops based on trends and projections.

3. Agricultural scientists depend on design of experiments and analysis of variance in the study of crop variations and soil fertility.

4. Biostatistics provides a lot of input to the research institutes such as Indian Institute of Chemical Technology (IICT), Central Drug Research Institute (CDRI), Indian Council of Medical Research (ICMR), Indian Council of Agricultural Research (ICAR) and others. For example, the study of growth and trends in the incidence of various diseases, growth and spread of various organs causing cancer etc. helps the scientist in the invention of new treatments and new drugs.

5. Pharmaceutical industry makes use of biostatistics in Drug discovery and development process, design of new drug delivery systems and formulation development, testing hypothesis in developing new drugs, formulations and for assessing the effectiveness of a drug in curing a disease.

6. The pharmaceutical industry also uses the tools of statistical control and theory of sampling for quality control and quality assurance of drugs and pharmaceuticals.

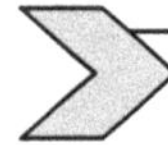 ## Specific Applications of Biostatistics in Pharmacy Research

(i) Provides methods of data collection and presentation and analysis of results for better understanding and for drawing valid conclusions.

(ii) During the stage of planning the experiments, design of experiments (DOE) provides inputs for planning experiments to get the relevant information.

(iii) Provide methods to reduce experimental error in the investigation

(iv) Provide methods to estimate sample size (n) or replication needed for the given accuracy in the results

(v) Provides standard designs and methods of analysis suitable for various experimental and observational studies.

(vi) Provides a variety of "tests of significance" to test the significance of the observed results and to make comparisons.

(vii) Provides methods of analysis of bivariate population and to evaluate the correlation between two variables.

(viii) Provides methods of regression and multiple regression for establishing mathematical relationship between two or more variables.

(ix) Provides methods of Probit analysis for estimating LD_{50} and ED_{50} of new drug substances.

(x) Provides statistical optimization techniques for drug development, formulation development and analytical method development.

CHAPTER 2

Research Methodologies - I
(Concepts, Literature Survey and Protocols)

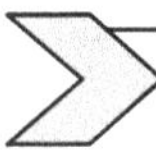

Research and Its Importance

Research is a systematic and organised scientific process to find answers to questions. For example a question is raised: Can olibanum resin be used as film former in film coating? How it is comparable to HPMC in film coating? Assuming these questions are not answered already in the whole of medical and pharmaceutical literature, they can be answered by conducting a scientific study. Such a study, if conducted is systematic because there is a definite set of procedures and steps to be followed in a specified order. It is organized in the sense that it is a planned process and not an impulsive one. The above questions are the basis and a starting point of research. If there is no question there is no research. Answer(s) to question(s) is the endpoint of research. The answer may be negative, but it is still an answer.

Research generates new knowledge that can be used to solve a problem or improve the existing status of a process. A pharmacist developing a new drug or a new novel drug delivery system, a

LEARNING OBJECTIVES

To know research and its significance Literature survey, sources and techniques

Research Protocols

♦ Importance of statistical methods in research.

♦ Thrust areas of research areas in pharmacy.

sociologist questioning villagers about their food habits, a statistician doing a modelling of pharmacokinetic data, a geneticist encoding a protein sequence- are all doing research though using different instruments to attain their objectives i.e., laboratory and instrumentation by the pharmacist, a pencil and paper by the sociologist, a computer by the statistician and molecular probes by the geneticist. Research uses scientific methods to discover facts and their interrelationships and the new knowledge obtained is applied in practical settings. Research provide a fresh understanding of a disease process, or mechanism of action of a drug or it may provide new tools for disease management such as vaccines or it may generate information on the health problems of a community to plan health care strategies.

Importance of Research

Knowledge is power and research is essential for generating new knowledge, information and understanding problems that can enable the community to achieve a better quality of life. In the context of pharmacy and medicine, research is undertaken to develop new drugs or therapies to promote health. There are four important reasons why research should be undertaken:

(i) *To promote basic knowledge*: This is the infrastructure upon which the drug treatment, disease management, or health care reforms depend.

(ii) *To develop new tools*: These may be drugs, vaccines, a diagnostic aid, pesticides an operative technique, an instrument or rating scales.

(iii) *For health promotion*: In recent years substantial improvements in health have resulted from changes in the life style, diet and activity – all of which are due to the outcomes of research.

(iv) *For effective planning*: Research provides the data for better management of scarce resources and can guide health policies and action.

The pursuit of research depends on systematic analyses, creativity, exploration and commitment to truth. The benefits of research, both social and monetary stimulate a demand for it and result in faster progress for the mutual benefit of society.

If we do research we ourselves can solve our problems; we can attain intellectual independence and stay in front and we can commercially exploit the fruits of research leading to economic growth. Research promotes the skills of independent thinking, reasoning, creativity and originality.

Research in PG and PhD Programmes in Pharmacy

Many Universities in India and abroad have made research work a part of PG curriculum. The aim is to teach a PG student the fundamentals of research methodology stimulate an interest in research. A PG student, at the end of the course, should be capable of planning and carrying out an independent research project. It also helps him/ her develop the necessary skills required to do research. Such skills will also assist him/ her to critically analyse literature and assess the relative merits and demerits of new concepts and their applications.

The reasons for doing a PG and PhD research are as follows:

1. *Learning research methodology*: PGs learn and get trained in research methodology.

2. *Development of scientific attitude*: A pharmacist must think scientifically and develop scientific attitude towards his/ her sphere of activity. Such attitude is useful in discharging their duties more efficiently.

3. *Chance for in depth study*: Research work offers the PGs an opportunity to study a topic in depth and earn experience in a particular field.

4. *Critical reading*: The PGs learn how to collect literature on a topic and analyse it critically instead of blindly accepting whatever is published. They get to know how to use the library and the internet for literature search.

5. *Special skills*: In the course of research, PGs may develop special skills and interests which they could put to good use in future.

6. *Contributing new information*: Contributing new knowledge, however small it may be, is exciting and satisfying.

7. *Publication*: Research work can be published in journals and as patents.

The Role of Research Guide

Research at PG and PhD level has to be done under the supervision of a research guide. The research guide is expected to supervise the candidate at all stages of the research work. From defining a problem to finally approving and signing the research work he/she should be involved in all stages. This does not mean that the guide should sit with the candidate and do the experiments or finally write the research report or thesis. The guide is expected to teach the PGs the basics of research methodology, cut through the red tape, smoothen administrative problems and help the candidate procure the necessary drugs or instruments and offers general help in the conduct of the study. The guide should also take it as a challenge to identify relevant areas of research, define challenging problems and ask the PGs to seek answers to problems of pharma industry, health care and society at large.

Qualities of an Ideal Research Guide

A research guide/ co- guide should be

1. Knowledgeable, competent and encouraging
2. Possess considerable interest in research
3. An expert in the field of interest
4. Have good communication and feedback
5. Available and approachable at all reasonable times
6. Solves administrative problems
7. Critical but flexible and listens to students
8. Courteous and respectful

The Role of Research Students

The research students should realize the importance of research and became truly involved in their research projects working sincerely to make the best use of the opportunity to learn research methodology, gain the practical knowledge and skills of organizing and conducting the work and have a firsthand experience of learning scientific writing - a skill which is required more and more in today's world. They should also insist on more involvement from their guides. Since it is customary for the PG and PhD students to be the first author of their publications which evolve from the research work, he/ she has to accept the responsibility for the same. It is probably the first time he/ she will be held accountable for their actions.

 **Literature Survey**

Literature survey is a systematic review of all scientific resources both published (i.e., journals, data bases and text books) and non published (i.e., registry, research and thesis) in order to gain information in an area of one's interest.

Literature survey is not a onetime task but a continuous process. It starts when a researcher is looking for a researchable problem in the area of his/ her interest and gets repeated throughout the study from writing the protocol to publishing the results so that he/she can update themselves about the information on the topic of interest.

Importance of Literature Survey

1. A review of literature will reveal which aspects related to a topic were researched and which were not. Unresearched topics can be chosen for the research work.

2. The pitfalls/weaknesses in others' studies can be identified and can be rectified in the proposed work

3. The difficulties faced by the other workers may be avoided by foreseeing or finding out solutions for some of them in advance. Some papers may offer solutions to the problems encountered in another study

4. Some workers may have described a modified procedure instead of the standard one. The modification may be the better one and may very well suit the purpose.

5. It can give an idea about the dose of drugs, animal models, number of subjects and statistical methods. These details can help the researcher design the study.

6. One can get an idea whether the study can be completed within the stipulated time period.

7. Review of literature also helps the authors write an effective discussion of their manuscript.

Sources for Literature Survey

Primary Sources

The journals are the primary source for a literature survey. Journal articles provide the latest information and help the researchers to update their knowledge in newer developments in the area of their interest apart from providing the details of design and methodology. Literature survey should be conducted in indexed as well as non indexed journals. The limitation is that all journals are not freely accessible.

Secondary Sources

Abstracting services like PubMed, EMBASE, Biosis and IndMed are the secondary sources. It is possible to do an effective and quick survey of the literature using the secondary sources. Abstract should never be used as a primary source since it can lead to misinterpretation. The abstracting services are useful for initial survey.

Every abstracting agency has a limited number of journals only and not all journals in the world. So it is essential to survey more than one abstracting agency. Yet another limitation is that the abstracts may not be available for all types of articles, i.e., editorial, research letter, case report and letters to the editor.

Tertiary Sources

Text books are the tertiary sources and are easily available. One should not hesitate to refer to all the available text books to find out what has been said about the chosen topic. The drug that is to be used in the study may be exhaustively coved in pharmacology text books. It would be wise to read not only the general text books but also the reference books. It is important that one should read the most recent edition of a book. It is also a good practise to read older editions which sometimes describe the historical back ground of a drug or procedure or a concept. The newer editions omit such information for want of space. Limitations of tertiary sources include the absence of the most recent developments as it takes several years to publish a new edition of the text book.

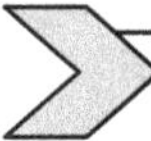 **Process of Literature Survey/Search**

The first step in literature survey is identifying the key words. For this purpose define the researchable problem as specifically as possible and break it into small topics. Think carefully about each topic and select all possible key words

The next is searching all sources. Once the list of keywords is made, start making a list of sources. Every effort should be made to search all the sources. The survey can start with the textbooks in the library. Once an article relevant to the topic is obtained, list of references given in the article should be carefully read and the important relevant references picked up and searched. This is one way of extending the survey. The survey then must be conducted on the internet. A cautionary note here is there are many layman websites, the sites maintain by individuals and the information given in them need not be reliable. The researcher should use his discretion before using or quoting such information. Literature survey on the internet should essentially include a search of the abstracting database like PubMed and websites of professional organizations, industry and Governmental bodies. Abstracting agencies give only the abstract and every effort should be made to get the full text of the article.

Literature Search Techniques

Searching by a Phrase

If the identified key word is a phrase, (e.g.) adverse drug reaction, the phrase should be mentioned within quotes while searching.

Boolean Searching

A search using Boolean operators is called Boolean searching. Once a list of key words for searching is ready, a way to link the key words together must be worked out. The linking is usually done by three terms- AND, OR, NOT. These terms are called Boolean operators. The Boolean operators are always written in the uppercase (full capitals). Boolean searching may help to narrow down or widen the search.

'AND' is used for narrowing the search. The search results should contain all the terms joined with this e.g. the search "asthma AND children" leaves out all the studies conducted in adult patients and narrows down the search to asthmatic children.

'OR' is used for broadening the search. The results contain any or all of the terms linked.

'NOT' is used to refine the search by narrowing it. The search "edema NOT heart failure" displays all studies on edema but NOT those due to heart failure.

Truncation and Wild Card Searching

A part of the keyword is replaced by a symbol, usually *. If the last part of the key word is replaced by * it is called as truncation searching. e.g. a truncation search with a keyword <u>sedat*</u> will carryout asearch on sedative and sedation. When letters inside a key word are replaced with a symbol then it is called a wildcard search. e.g. the keyword <u>am*biasis</u>will look for both amoebiasis and amebiasis.

Focused Searching

The search can be focussed or refined by giving limits to the search. A few ways of limiting the search are to include the following limits in the search:

- Experimental/clinical study
- Age, gender of the participants of the study
- Design of study- randomized controlled trail, observational study
- Type of publication – meta analysis, review, original research, case report
- Date of publication

After Initial Search

Once the initial search in all the sources is over, record the date of search, key terms used, methods followed and results obtained. Carefully go through the results obtained for relevance, quality and quantity. All the relevant results should be retained and saved. If it is a text book, necessary pages along with the first two pages (where one can find the details of the publisher) should be photocopied and filed. If it is a chapter in a book, the whole chapter should be photocopied and filed. All the materials saved in electronic format (PDF of articles, Web pages) should preferably be also printed out and filed.

Start Reading

After filing the search results, one should start reading them. This will certainly improve the knowledge on the topic which will help the researcher to plan and conduct the study in a better way.

Keep Searching

Literature search is a dynamic process and continues throughout the study. It should be carried at frequent intervals. This is one way by which a researcher can keep in touch with the recent developments in the topic of interest.

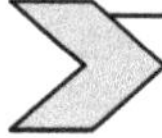 # Research Protocol

A research protocol is a well organized framework of instructions and procedures to be implemented in a systematic manner for the conduct of research. Research protocol by convention means a preliminary draft in which the information and/or knowledge needed for the actual conduct of an experiment is compiled briefly to outline the execution of the study.

Need for a Research Protocol

A clear and concise protocol guides the researcher step by step throughout the study and also provides solutions to problems that may arise during the course of the research project. Ideally the protocol should be prepared in consultation with the guide to incorporate his/her suggestions or modifications. This helps in plugging loopholes in the plan of study.

A good protocol can serve as a framework when writing the manuscript after completing the research. It keeps the researcher on track by efficient management of time and resources. It also allows the Institute Research Council (IRC), Institute Ethics Committee (IEC) and funding agencies to judge the proposed research work.

Hence a protocol is an essential prerequisite for any scientific research and no research work should be conducted before finalizing the protocol.

Elements of a Research Protocol

The components of a protocol can be summarized as follows:

1. Title
2. Names and affiliations of all investigators with contact details (phone number, email ID)
3. Introduction the actual research problem and justification
4. Aims and objectives
5. Hypothesis
6. Review of literature
7. Methodology
   - Study design
   - *Population*: Inclusion and exclusion criteria
   - *Intervention*: Details of intervention/drug (formulation, dose, frequency of administration), duration of intervention etc.
   - *Control*: Control (Standard Drug or Placebo)
   - Size of the samples in each group
   - *Outcome*: Primary (efficacy), secondary (safety, ADRs) parameters to be measured
8. Brief procedure
9. Statistics
   - Sample size calculation
   - Statistical test for each parameter
   - Level of significance
   - Method of analysis
10. Ethical issues
11. References
12. Signatures of the investigator, guide, co-guide and Head of the Department.

Statistics and Its Importance in Research

The term statistics is often used to indicate facts and figures of any kind such as health statistics, vital statistics, business statistics etc. it is also used to refer to a body of knowledge known as statistical methods developed for handling data in general, particularly in the fields of experimentation and research. Statistics is more precisely defined as the science of collecting, summarizing, presenting and interpreting data and of using them to test hypothesis. The science of statistics provides principles and methods for collection, presentation, analysis and interpretation of numerical data of different types such as observational data, qualitative data, data obtained by a repetitive operation etc. Statistics is essentially concern with study of chance variation or variability. Chance variation is the variability that arises in the data due to factors or reasons beyond our control. Chance variation leads to 'uncertainty' and valid conclusions cannot be drawn from the data. Statistical methods and techniques are helpful in this situation to eliminate variability or uncertainty in the data and to draw valid conclusions.

Statistical methods are useful for the following purposes in experimental methods and research.

1. To reduce the variation in experimental material data

2. To provide valid inferences, based on which one can draw reasonable/ valid conclusions from a study.

3. For making comparisons i.e., to compare the effectiveness of one drug with that of another etc.

Thrust Areas of Research in Pharmacy

1. Drug discovery and development
2. Formulation Development (Generic and Branded)
3. Preformulation Studies
4. Novel Drug Delivery Systems (NDDS)
5. Bioavailability and Bioequivalence
6. Biologics and Biosimilars
7. Nano Technology and Nano Medicines
8. Biotechnology
9. Protein and Peptide Drugs
10. Pilot Plant and Scale up studies
11. Analytical method development and validation
12. Impurity profiles and methods for impurities and related substances
13. Pharmacokinetics and Drug metabolism
14. Drug Interactions
15. Phytochemistry and Natural Products
16. Herbal product Development
17. Standardization of Herbal Drugs
18. Pharmacological Screening methods
19. Tissue and Cell cultures
20. Pharmacodynamic evaluation
21. Toxicicological studies

22. Clinical Studies
23. (Phase I, II, III studies)
24. Pharmacovigilance
25. Pharmacy practice and hospital pharmacy studies
26. Regulatory Aspects
27. Nutraceuticals
28. Testing and analysis of foods and cosmetics
29. Formulation development and evaluation of cosmetics
30. Development of drugs for tropical diseases
31. Development of drugs for communicable diseases

Chapter 3

Research Methodologies - II
(Designs for Preclinical and Clinical Studies)

The study design should answer all the research questions raised in the aims and objectives. If the study design is found to be wrong at later date, the project cannot be salvaged and the entire effort goes wastage.

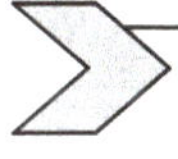

Criteria of a Study Design

A study design should satisfy the following criteria:

1. It should be able to test the research hypothesis properly.

2. It should control for extraneous variables like bias, confounders and sampling errors.

3. It should allow generalisation of results of the study to the population.

While designing the research all these three criteria should be taken in to account. It should be noted that sometimes, there is no single correct design for the hypothesis and more than one design may be appropriate for the hypothesis.

LEARNING OBJECTIVES

To know types of study designs Observational, descriptive and experimental designs

♦ Applications of various designs
♦ Errors and biases
♦ Strategies to eliminate errors and biases
♦ Error and its measurements
♦ Purpose of methodology
♦ Specific protocols for various types of studies (Animal, laboratory based, observational, descriptive, epidemiological and clinical studies)
♦ Sample size (n) and its determination.

⟩ Types of Study Designs

Study designs can be broadly classified into either observational or experimental.

Observational Studies

The subjects are just observed for the parameters to be measured and no active intervention is done. Based on the time frame, the observational studies can be divided into cross sectional and longitudinal .The longitudinal studies are further divided into case control (retrospective) and cohort (prospective) studies. The observational studies can also be classified as descriptive study, where the characteristics are described and analytical study, where analysis is carried out to ascertain an association between two factors.

Descriptive Study

A descriptive study is done to describe the various characteristics of the study subjects without causation or hypothesis testing. A good descriptive study should answer the questions like who, when, what, where, why and more importantly "so what". An important point should be brone in mind while designing a descriptive study is the need for a specific and stringent criteria for case definition. For example, the descriptive study is designed to describe the characteristics of the distribution of a rare disease in relation to Person – age, gender, occupation, personal/socioeconomic characteristics Place- geographic variation Time – seasonal/diurnal variation.

Types of Descriptive Studies

Descriptive studies can be classified based on whether the study deals with individuals or population. Individuals are observed in case report, case series and cross sectional study where as an ecological correlational study deals with population.

Case Report

This is a descriptive study of an individual (a single case) who has encountered a rare event. This forms the basis of a more detailed study. This is the least publishable unit in medical and pharmaceutical literature.

Case Series

Description of a number of individuals (many cases) with the same disease/event is called case series. The clustering of cases can be a signal for a cause- effect relationship and necessitates further observational/ experimental studies. For example, in 1961, a series of cases of phocomelia were reported in children born to mothers who had consumed thalidomide during their pregnancy. These case reports formed a case series which was a signal for the association between thalidomide and phocomelia. This led to designing of case control studies which proved the causal association of thalidomide to phocomelia.

Cross Sectional Study

It involves the study of a random sample of a population or a group of patients at a single time point. Cross sectional studies are the best design to study the prevalence of a disease. That is why they are also called as prevalence studies. These studies can identify the health needs / attitudes and help in planning and implementation of various health programmes.

For example, a cross sectional study on malnutrition may wish to establish percentage of malnourished children; socio economic, physical, political variables that influence the availability of food; feeding practices; knowledge, beliefs and opinions that influence these practices. By comparing malnourished and well nourished children, the study can determine which independent variable (i.e., socio economic, behavioural) contributed to malnutrition.

Ecological Correlation Study

This is called ecological study because the study is at the level of the entire population and not at the individual level. In this type of study, the suspected risk factor is measured in various populations and compared with the incidence of the disease to generate a hypothesis about the risk factor and disease. An important limitation of this design is that it can establish association only and not a causal relationship. For example, the number of safe deliveries is related to regular antenatal visits by the mother.

Application and Uses of Descriptive Studies

A descriptive study can be under taken when the disease is new or rare and much is not known about the etiology, symptoms and disease characteristics. Descriptive study is the first approach to research in a new event. For example, in 1990s SARS was virtually unknown and the initial descriptive studies on the symptomatology and presentation paved way for further research.

Descriptive studies provide data regarding the magnitude and types of disease in the community; provide clues to disease etiology and help in formulation of an etiological hypothesis; provide background data for planning, organising and evaluating preventive and curative services and used to describe variations in disease occurrence by time, place and person.

The major limitation of descriptive study is that it does not account for chance variation. For example, the improvement or worsening of a disease can be due to the fluctuating course of the disease or due to chance/ inter-individual variation. This cannot be ruled out as there is no control group also it cannot account bias and confounders.

Case Control Study

A group of patients with disease (cases) is compared with a group of individuals without disease (control). Both the groups are followed backwards in time to determine the association with the risk factor. Hence this is also called retrospective study.

Example: Pregnant women at term with (a) foetal malformation and (b) normal foetus are selected for the study and traced back to find out whether they took a particular drug during the first trimester.

Selection of Cases and Controls

The case and the source population (population from which cases arise) should be well defined. Previously existing cases are not preferred, as diagnostic criteria changes with time and the exposure may affect the prognosis and duration of the disease. The validity of a case control study depends on the selection of controls. They should be selected from the same population as cases and it should be independent of exposure. Increasing in the number of controls increases the power of the study.

The data gatherers should not be aware of the status of the participants (case/control) and the study hypothesis. As the cases and controls differ in recollecting the exposure, the data gatherers should be trained to elicit history of exposure in a similar manner in cases and controls. This helps to eliminate bias.

Applications of Case Control Study

1. It is an efficient design for the study of rare diseases.
2. It can be employed for studies in diseases with long latency period like AIDS, Cancer.

Advantages

The study is quick, cheap and easy to conduct; requires smaller sample size than other designs and many risk factors can be investigated.

Cohort Study

In this type of study design, the researcher assembles a group of people (cohort) and follows them forward in time. The exposure status to a risk factor and the outcomes are observed in this cohort. In Roman military, a group of 300-600 soldiers marching forward was called a cohort. As this study proceeds forward from exposure to outcome, it was named as cohort study.

In general, all cohort studies are prospective. However there is a study called 'retrospective cohort' in which the researchers goes back in time to form a group of people from the case records/registry and observes them for outcomes. In this type of study also, the direction of observation is always forwards.

Cohort study is susceptible to selection bias. For example in a cohort to determine the effect of jogging on cardiovascular events, the researcher should be careful in selecting the subjects as people who jog are more health conscious and their life style and dietary habits may also differ. The outcome should be defined clearly and the person/researcher who measures the outcome should be unaware of the status of exposure of the subjects (blinding).

Cohort study is the best design to establish the natural history and incidence of a disease. Relative risk can also be calculated. Establishes the time sequence of events (temporal association) and this strengthens the inference. Multiple outcomes of a particular exposure can be studied.

The major limitation is time consuming and costly. It requires large sample size and hence not suitable for study of rare diseases.

Experimental Studies

In experimental studies an active intervention is done and the subjects are observed for its effect. An experimental study should satisfy the following criteria:

1. There should be an intervention (i.e drug therapy/ exercise/ surgery)
2. It should include a control group without intervention.
3. The participants/ animals should be randomly assigned to different groups.

If one or more criteria are not satisfied, then it is known as quasi experimental study. Experimental studies can be broadly divided into animal studies and human studies.

Animal Studies

Animal studies can be *in vitro* (experiment or reaction occurring outside the body in a test tube/ culture medium), *in vivo* (experiments on intact animals) or *ex vivo* (tissues/ cells are removed from the animal and experiments are carried out on it by maintaining it in a viable state).

Human Studies

Experiments carried out on humans are called clinical trials.

Clinical Trials

Clinical trials are carefully designed experiments to answer specific questions or to test a hypothesis. It consists of a group subjected to an intervention/ drug and the other group as a control. It can be non randomised or randomised. Randomised control trials (RCTs) are the standard designs to evaluate an intervention/ treatment.

(i) *Parallel Design*: In this design two separate groups are treated concurrently by two different ways (standard drug/ new drug). This design is commonly used.

(ii) *Paired Design*: Only one group is used (say n = 10). After measuring the baseline values in all subjects, an intervention (e.g. drug) is introduced. After giving sufficient time for the intervention to act, measurements are taken again and the data obtained before and after the intervention are compared.

(iii) *Cross Over Design*: Here one half of the subjects will receive the active treatment (e.g., drug) and the other half control treatment (e.g., placebo). After a washout period, the treatments will be crossed over so that all the subjects will receive both the treatments in sequence. This design takes care of not only the inter-subject variation but also the time effect and the carryover effect.

Error and Biases

Errors in measurements can occur in any study and they can adversely affect the outcome. These errors may be random or systematic. The systematic error is called as bias and random error is simply known as error. Random error is two sided and bias is one sided. For example the readings of a balance show ±0.1 g of real weight of the objects weighed, where as another balance shows 2 g more than the real weight. The first one is error can be on the positive or negative side. The other balance is biased since the error is always one sided. Errors and biases must be identified and eliminated as far as possible or at least controlled to minimize their effects so that validity of the study is preserved.

Strategies to Eliminate Errors and Biases

1. *Controls*: In many experimental designs, a control group is a must. The control group must, as far as possible, resemble the test group of subjects under investigation in all respects, except for the factor or the variable being investigated. Only then, comparisons can be made to conclude whether the intervention (i.e., drug) affects the groups

differently. The control subjects must match the test subjects in all aspects e.g., age, weight, and gender distribution should not be different in both groups. In some studies there may not be any need for a separate control group and the test group itself can act as its own control. For example in a study to evaluate the effects of topical antifungal agents on superficial functional infections of the skin, the corresponding site on the other side of the body can be used as a control. Historical controls are subjects whose data are obtained from old records and used as control values for subjects being treated now.

2. *Randomization*: Randomization (or random allocation) is the process by which every participant has an equal chance of getting any intervention. Hence randomization ensures that the participant get the intervention purely by chance, not choice. There are two important reasons for advocating randomization. It helps in distributing the various characteristics (which can affect the outcome) randomly between the groups. This reduces bias. Secondly, most of the statistical tests based on the assumption of random sampling. Randomization is done by random number number tablets or computer generated random numbers. Software packages like Rando and Random Allocation can be used for this purpose.

3. *Allocation Concealment:* The allocation sequence (randomization sequence) generated should be concealed i.e., the investigator/ recruiter should not know which intervention the next participant is going to get/which group the next participant is going to be allotted to.

4. *Placebo*: One has to make sure that placebo effect does not effect the conclusions drawn. Control group is given placebo and the effect it produces is compared with that in the test group so that the real effect of a drug or a treatment can be measured.

In animal studies, drug solutions are usually freshly prepared prior to the experiment. If the drug is dissolved in distilled water or propylene glycol, then the same vehicle should be used for the control animals.

5. *Blinding*: Blinding or masking is a process to make patient and/or researcher unaware of the nature of intervention (standard drug/new drug) provided to the participant. It can be single blind (either the patient or the investigator alone is blinded) or double blind (both patent and investigator are blinded) or triple blind (patient, investigator and the person who analyses the data are blinded). Blinding eliminates ascertainment bias.

To make blinding effective, the placebo or standard drug is made to resemble the test drug (shape, size, colour, weight and odour). If both the interventions were given by different routes, then a 'double dummy technique' can be followed. For example, if a new oral drug has to be compared with the standard parenteral drug, then the group receiving new oral drug will also receive a parenteral injection (placebo) similar to the standard drug. The group receiving the parenteral standard drug will also receive a placebo tablet which is similar to the new oral drug.

6. *Standardisation*: The experiment/ the procedures /the analytical techniques have to be standardized so that the results are reproducible. The process is repeatedly carried out by the researcher till he/she gets reproducible results. This reduces error and bias if any.

7. *Replication*: Replication reduces errors. The measurements are made repeatedly for a suitable number of times (n) and the mean value is used.

8. **Study *Design***: Some study designs inherently take care of error and bias. Crossover design and paired design are example of such designs.

Error and Its Measurement

Error is defined as the difference between true value and the estimated value. True value is related to the underlined population. For example the labelled drug content of a tablet or syrup is an example of true value. The tablet or syrup is assayed to know the actual amount of drug present in them. The preparation is generally assayed a suitable number of times, say 6 times and the average of the six assays is taken as the estimated value of the drug content. In such a case the error involved is measured by calculating Percent error as follows:

$$\text{Error (\%)} = \left\{ \frac{\text{estimated value} - \text{true value}}{\text{true value}} \right\} \times 100$$

In pharmacopoeias ±5% error is allowed for the content of active ingredient for pharmaceutical preparations in their monographs. The error may be positive (higher) or negative (lower).

In experimental investigations the true value is unknown and physically not existing. It is a hypothetical value. To know the unknown true value an experiment has to conducted and the result is taken as an estimate of unknown true value. For example in a study on evaluation of the antidiabetic activity of a herbal extract the true value is all possible measurements of the antidiabetic activity after administering the extract. If the measurement of the antidiabetic activity is done in six subjects or animals we will get six measurements of the antidiabetic activity. The average of six measurements $(\bar{x})$ is taken as the estimated value of the antidiabetic activity of the extract. The estimated average $(\bar{x})$ is not exactly equal to true unknown value. It involves error. The error in this case is the difference between the unknown true value (μ) and the estimated value $(\bar{x})$ The error in the estimated average is measured by estimating standard error (s.e.). the standard error (s.e.) is defined as

$$s.e = \frac{\sigma}{\sqrt{n}}$$

where σ is the standard deviation of the population and n is the sample size or number of observations in the sample. When σ is not known it is substituted by standard deviation of the sampled observations (sd). The six observations obtained by conducting the experiment six times as considered as a random sample taken from the unknown underlined population. In such cases the error involved in the average estimates is calculated as follows

$$s.e = \frac{sd}{\sqrt{n-1}} \quad \text{for a small sample}$$

In experimental investigations the samples obtained are always small only having less than 30 observations. A sample with less than 30 observations or data is known as a small sample in statistics.

Factors that contribute to the experimental error include are as follows:

1. *Chance factors*: These are minute changes in the experimental conditions such as temperature, pH, volume of reagents added etc. Alone these factors cannot produce a measurable change in the response, but collectively they bring out a measurable change in the response. These factors are beyond the control of experimental investigator. Their occurrence is also random. They may appear in one run and may not appear in other. The chance factors are mainly responsible for the differences observed in the response of the experiment when it is repeated.

2. *Assignable factors*: These factors can be identified and hence can be corrected to reduce the error. Some of these factors are as follows.

    (i) Variations in the quality and purity of materials (drugs, excipients, solvents, etc.) used in different runs. The error due to this factor can be reduced by using materials with the same specifications or quality in all the runs. For this purpose materials of highest purity grade are to be used in research investigations. In the case of excipients specifications related to particle size, molecular weight, viscosity etc. are to be defined initially and excipients conforming to the required specifications are to be used in research investigations. The drug, excipient and other materials are to be tested and confirmed that they are of good quality with regard to the specifications laid down before they are used in research investigations.

(ii) Working status of instruments and equipments. The instruments and the equipments used in research investigations are to be in good working condition for reliable and reproducible results. All the instruments and equipments are to be calibrated and validated periodically to ensure that they are in good working condition before using them in research.

(iii) Variations in methods and techniques used. Variations in the methods and techniques used contribute to a large error in the results. All the methods and techniques are to be standardised and validated for accuracy and precession and are to be used as per a defined SOP.

(iv) Improper use of instruments and methods will contribute to a large error. The investigator should be well versed with operating procedures of the instruments and methods. The investigator should acquire necessary skills.

(v) Inadequate experimental design and replicates will also contribute to error. Appropriate design and using the principles of randomization, local control, crossover with adequate sample size will reduce error. All efforts should be made to reduce error minimum. Small error enables us to make sensible comparisons.

Methodologies and Protocols

The purpose of methodology is to permit any person who wants to duplicates the study to do so. It should be taken like a recipe, carefully with attention to detail giving all the specifics of the drugs given, instruments used, experiments or interventions done and the statistical analysis. Use subheadings wherever needed, for clarity.

Provide as much details as necessary for a person to read what is written and conduct the study without seeking any clarification. For standard procedures (such as measuring blood pressure using a mercury sphygmomanometer) and well established laboratory tests, for example, like the glucose oxidase method for estimating blood sugar, there is no need to describe the method in detail. It is enough to give the reference for the method. However, if the method was modified and the modified version was used in the work, then describe the whole method. Give references for all uncommon and relatively new methods (including modified standard methods) which were used. There is no hard and fast rule about what needs a reference and it is something one will learn by practice.

Specific Protocols for Various Types of Studies

Details to be given in the protocols of various types of studies are as follows:

I.   *Animal Studies*: studies which use animals and are usually of experimental in nature.

1. *Animals*: species, strain, number, age, gender, special types (pregnant/weanling), wait, rationale for selection.

2. *Housing conditions*: single/group, if group how many to a cage, light/dark cycle, food and water, temperature, acclimatization (if done-how long).

3. *Groups*: name of the group, how many to a group, what drugs/interventions done to each group.

4. Drugs-name (generic), dose per kg body weight, number of doses per day, route, timing of dose

   - Name of the company where drug was produced, how the solution is prepared. (fresh/stock solution), vehicle for dissolving, storage conditions of stock, drug

   - If plant extract was used- give full name of plant, which part of the plant was used, when collection was made, how was it extracted, which extract was used for study.

5. *Surgical procedure*: method, anaesthesia, mortality, sham operation for controls.

6. *Tests*: Describe method completely (if well established method, give reference e.g., testing for analgesia using tail flick assay in rats)

7. *Instrument*: name, company, software (name, version, company).

8. *Ethics clearance*: animal ethics committee obtained or not.

9. Statist*ical analys*is: tests used, if more than one test say which test was used for which parameter, level of significance.

II.   *Laboratory Based Studies*: Those studies which are mainly done using samples of body fluids, tissues etc., (biochemical, micro biological, pharmacokinetic studies).

1. Specimens collected from whom, what time, how many samples, method of collection, how were they stored (or analyzed immediately).

2. *Method of analysis*: name, instrument for measurement, company.

3. *Describe method of analysis*: if kits were used give details of name of kit, company, sensitivity, whether duplicate or triplicate samples were used.

4. If other parameters were calculated using the plasma levels or serum levels mention which formulae were used. If dedicated software was used give the name/version/company.

5. *Statistical analysis*: tests used, if more than one test say which test was used for which parameter, levels of significance.

III. *Descriptive studies*: Clinical studies which involve a description of the clinical presentation and laboratory finding in a series of cases sometimes referred as clinic- pathological descriptive studies.

1. Setting – hospital OPD/in –patients, primary health centre, sub centre etc.,

2. Time during which data collection was done (give months and years of the starting and stopping time), was it prospective or retrospective.

3. Ethics clearance

4. Description of cases selected, diagnostic criteria, inclusion criteria, exclusion criteria, whether cases were divided into groups – if so what was the criteria used for the grouping

5. What was done – tests performed, normal/ abnormal values, treatments – medical/surgical procedure

6. How was out come measured – criteria, measured after how long, who assessed outcomes

7. Statistical analysis – were all cases included, if any were not included give reason, tests used, compared which group with what, level of significance, confidence intervals

IV. *Epidemiological Studies*: Community based studies, pharmaco-epidemiological studies, surveys as well as other designs.

1. Detailed study design; prospective/ retrospective, survey/ case control/cohort

2. Description of the population studied: catchment area

3. Sampling frame

4. Sample size (how obtained)

5. Method of sampling: describe the method used – random, snow-ball, purposive, cluster, stratified random etc.,

6. Ethics clearance, informed consent, was community clearance obtained?

7. Instrument or tool used- Questinnaire – if previously published scale/Questinnaire used give reference, was a translation used? Was it pre – tested? Was the Questinnaire validated? Give full Questinnaire. who administered the Questinnaire? Were they trained? Where was it administered (in the houses/school/community room)

8. If any measurement was done using an instrument(s) (e.g. measuring blood pressure with sphygmomanometer/ how frequently were the instruments calibrated, was there a difference between the instruments used (if more than one instrument), inter – rater variability (inter individual), intra individual variability, mention whether this were checked

9. Statistical analysis – spread sheet used for data entry, was data entry checked?, software used for analysis, tests used, if more than one test, say which test was used for which parameter, level of significance

V. *Clinical Trials*

1. *Setting*: place, level of care
2. Planned study population
3. *Inclusion criteria*: definite terms e.g., hypertension (in the study what did you consider as hypertension)
4. Exclusion criteria
5. Planned interventions and their timing
6. Ethics clearance/informed consent
7. *Sample size*: how calculated/how target samples size was protected
8. *Sampling procedures:* random allocation (samples/stratified/block/geographical) done or not, if so how was it done-computerized/ tables, controlled by whom
9. Method of allocation concealment and timing of assignment, method to separate the generator from the executor of assignment
10. *Masking (Binding)*: Describe mechanism (e.g., capsules, tablets), similarity of treatment characteristics (appearance/taste), allocation schedule control evidence for successful blinding among participants, person doing intervention, outcome assessors and data analysis
11. *Outcome measures*: primary and secondary outcome measures and the minimum important differences
12. *Dropouts*: number, causes, at which stage
13. *Analysis*: Rationale and methods for statistical analysis

Determination of Sample Size

An important item considered while planning any investigation is the number of cases required. It is unethical to include too many cases in an experiment. Including too many cases in a study or an experiment may also increases the cost of research. On the other hand, studying to small a number of cases may result in finding that may not be significant. There is no simple answer for this frequently asked question on sample size needed. Taking in to account the nature of the measurement of the observation, the design of the investigation and variability in the characteristic in the population, the number of cases to be studied, the sample size, is calculated.

Specifications needed to Calculate Sample Size

To estimate the sample size needed for a study or an experiment the statistician or investigator should know some important aspects about the study. They are:

1. The precision with which the parameters should be estimated or differences identified; or the minimum difference between parameters which is sufficient clinical importance.

2. The degree of safeguard allowed against the risk of drawing wrong conclusions (confident level). This implies controlling Type I (α) and Type II error (β). Often Type 1 error is considered.

3. An approximate idea of the population value, or the degree of variability.

In general, the more the variability, the precision and confident level, the larger the number of cases to be studied in specific in specified circumstances.

Examples of sample size estimation

1. For estimating Mean

   Sometimes, the investigator is interested in estimating population mean. For determining sample size, the investigator needs to know the variation in the variable under consideration and the expected difference between the sample mean and the population that can be tolerated. If σ is the standard deviation, that represents the variation and *d is* the difference tolerated, then with 95% confidence interval the sample size needed is given by

$$\text{Sample size } n = Z^2.\sigma^2/d^2$$

   Where Z is standard normal variate corresponding 95% confidence

Example: An investigator wanted to estimate the mean birth weight in a population, the standard deviation from a sample was known to be 450 gm. The expected difference between the population mean and sample mean was considered to be 30 gm. Then the sample size n needed is

$$\text{Sample size } n = Z^2 . \sigma^2/d^2 = 1.96^2 \times 450^2/30^2 = 864$$

The sample size needed is 864.

The sample size in testing difference in mean

To determination the sample size n, the following formula is

$$n = 2 \times [Z_\alpha \, (\sigma)/d]^2$$

If, in addition to α we consider the specific power $1 - \beta$, the sample size is determined by the formula:

$$n = 2(Z_\alpha + Z_\beta)^2 \, \sigma^2/d^2$$

Example: In the study of serum uric acid levels, the mean level in the diseased group is 5.4 mg/100ml. Using a new drug, the investigator expects the clinical difference between the untreated and the treated group to be at least   0.2 mg/100ml. The investigator wishes to conduct the experiment with the significance level $\alpha = 0.05$ and power $(1 - \beta) = 0.90$. Then the appropriate sample size needed for each group is

$$n = 2 \times [(1.96 -(-1.282)) \times (1.1)/0.2]^2 = 636$$

Supposing the investigator specifies the significance level only, then the sample size needed each group

$$n = 2 \times [(1.96) \, (1.1/)]^2 = 232$$

Alternatively sample size $n = 2(Z_\alpha + Z_\beta)^2 \, \sigma^2/d^2$ for each group or arm

The investigator wants a 90% power $(1 - \beta)$ of the test, so $Z_\alpha = 1.96$, Z_β 1.282. $\sigma^2 = (1.1)^2$,

d = difference between the mean expected = 0.2.

Sample size $n = 2(1.96 + 1.282)^2 \times (1.1)^2/ 0.2^2 = 636$ for each group or arm

For Estimating a Proportion

Example: Suppose the attack rate of disease in a locality is 5% (from past record) (i) The investigator allows an error of 20% of the attack rate on either side (i.e., the investigator wishes to get his estimated value between 4% and 6%) (ii) The investigator wishes to correct in his estimate (any value between 4 and 6) in 95 out of 100 attempts (safety factor 5% i.e., plus or minus standard error distances, at which the chances are 20 to 1 against exceeding that range of error).

Sample size $n = 4pq/d^2$

Where $p = 5$, $q = 95$ and $d =$ difference tolerated $= 20\%$ of p ($=1$) and $4 = 2^2$ (2 is the approximation of 1.96 which is the 95% confidence limit of the normal distribution.

$$n = 4 \times 5 \times 95/(1)^2 = 1900$$

For Testing difference between Two Proportions

An investigator want to assess the efficacy of a new drug or vaccine. If the investigator feels that the vaccine will be of value only if the produces a protective effect of 50% of the control (for example, if 20 cases of the disease in question are expected among 100 controls, then only 10 cases are expected among the vaccinated), then the required sample size in the study can be determined by

Without Considering the Power of the Test

Sample size $n = 2\,Z^2 pq/d^2$ for each group or arm

Where $p = 15 = (P_1 + P_2)$, $q = 85$, $d =$ difference tolerated $= 10$, i.e., $(P_1 - P_2)$ and $= Z_\alpha$ is 1.96 which is the critical value at 5% level of the normal distribution rounded off as 2.

$$n = 2 \times 2^2 \times 15 \times 85/10^2 = 100$$

Considering the Power of the Test

Sample size $n = 2(Z_\alpha + Z_\beta)^2 pq/d^2$ for each group or arm

If the investigator additionally states that a 80% power $(1 - \beta)$ of the test is needed, then $Z_\alpha = 1.96$, $Z_\beta = 0.842$, $p = 15$ and $q = 85$.

Sample size $n = 2(1.96 + 0.842)^2 \times 15 \times 85/(10)^2 = 200$ for each group or arm.

Chapter 4

Basic Concepts of Biostatistics

Statistics and biostatistics concerns with collection of data, presentation and analysis of data, testing and interpretation of data to draw valid conclusions. Data forms the raw materials for all statistical methods. Data is a collection of observations expressed in numerical figures. The data are collected in two ways (i) by conducting an experiment and (ii) by making a survey. In pharmacy and scientific research the data are collected by conducting experiments. Where as in social sciences including pharmacy practice research the data are collected by making a survey.

LEARNING OBJECTIVES

- Data and variables, types.
- Population and sample, random sample, concept of probability.
- Presentation and analysis of data in tabular form, graphs, curves and pictures.
- Frequency distribution table, frequency distribution curves, histograms, bar diagram and pie diagram.

Types of Data

The statistical data can be divided into two categories:

1. Qualitative (attributes or characteristics)
2. Quantitative

Qualitative

In this type of data, there is no numerical relation with one another.

Examples: skin colour – brown, black and White.

Eye colour – blue and brown.

Sterile or non- sterile.

Defective or non defective items.

These data are also called discontinuous variables or attributes.

Quantitative

1. In this type of data, there is numerical relation with one another.
2. It may be continues or discrete.

Example: Discrete -equal to no.of books, no.of students.

Continuous – height or weight of person.

These data are also called continous variables. In pharmacy and research investigations the data are of continous type only. The following are some examples of continous variables in research/pharmacy

1. Yield of a chemical reaction in chemical experiments
2. Globule size in emulsion in pharmaceutics
3. Blood gluclose levels in pharmacokinetics
4. Weight of a patient in hospital pharmacy
5. Zone of inhibition in microbiological assays
6. Height of contraction of the muscle when a drug solution is added in a typical pharmacology experiment.

The properties of qualitative and quantitative data are compared in table 4.1

Table 4.1 Comparison of qualitative and quantitative data.

Qualitative	Quantitative
1. Always discrete.	1. Discrete or continuous.
2. No magnitude.	2. Have magnitude.
3. Persons with same character are counted to form groups.	3. Arranged by both character and frequency.
4. Results are expressed as ratio or proportion.	4. Such data are analysed by statistical methods.

Classification of Data

A. According to source of data collection:
   (i)   *Primary data*: directly from field or experiment.
   (ii) *Secondary data*: obtained from primary data or review.

B. According to variable:
   (i)   Univariable.
   (ii) Bivariable.
   (iii) Multivariable.

C. According to population:

    (i)   *Raw data*: Data from population.

    (ii) *Derived data*: Calculated from primary value of data.

Primary Data

This data are collected directly from the field of enquiry for a specific purpose, these are raw data or data in original nature and directly collected from population. The collection of primary data may be made through either by complete enumeration or sampling serve method

Secondary Data

These are numerical information which have been already collected by some agency for a specific purpose and are subsequent compiled from that source for application in different collection.

In other words, data used by any other agency, then the collection authority will be termed as secondary data.

Collection of Primary Data

The following methods are generally used for collection of primary data:

    (a) Direct personal observation.

    (b) Indirect oral investigation.

    (c) Questionnaires sent by mail.

    (d) Schedules sent through investigators.

Questionnaire

It is a proforma containing a sequence of questions relevant to a statistical enquiry. It is used for collection of primary data from individual persons through their response to the set of questions.

Relative Advantages of Primary Data

1. Primary data provides with detailed information but in secondary data some information may be supressed.

2. Primary data is free from transcribing errors and estimating errors where as a secondary may contain such errors.

3. Secondary data normally do not contain information regarding methods of procuring data whereas primary data often include them.

4. Cost effectiveness is a vital plus point for using secondary data.

Thus time, cost suitability and accuracy are the essential factors whether we would use primary or secondary data.

Population and Sample

Population is an entire group of people or study elements like persons, things or measurements having some common fundamental characteristics. Population may be physically existing or hypothetical. Examples of population that physically exist include a batch of tablets produced, 100 litres of a syrup manufactured, number of patients in a hospital , number of students in a college etc.

In scientific investigation and research several times the population is hypothetical and does not physically exist. Examples include

(i)   *Yield of a chemical reaction*: All possible measurements given as yield of a chemical reaction from population for investigation on chemical reactions.

(ii) *Haemoglobin content of normal male adults*: All possible measurements given as haemoglobin content of normal male adults.

(iii) *Anti diabetic activity of a herbal drug*: All possible measurements given as anti diabetic activity (percentage reduction in blood glucose levels).

In the above three examples the population is not physically existing. The population is hypothetical. To obtain the data values in the population an experiment as to be conducted. If the experiment is conducted ten times (n=10) we will get ten data points of the population. These ten data points form a sample obtained from the hypothetical population. The sample data is then analysed and interpreted to take a decision regarding the population.

Sample

Sample is a part of population or a small group of selected subjects or items drawn from a population for studying the characteristics of the population. In cases where the population is large like tablets produced in a batch, a sample is taken and studied instead of studying the entire population. The result of the sample is extended to the population and a decision is taken about the population.

For example a batch of ten lakhs paracetmol tablets are produced. The ten lakh paracetmol tablets form the population. For quality control purpose a sample of 50-100 tablets are taken and subjected to quality control testing. Based on the results of the quality control testing of the sample a decision is taken about the population. If the sample of tablets fulfills all the quality control specifications prescribed, the population i.e., the ten lakh batch of paracetmol tablets prepared is accepted.

Types of Samples

(i)   *Random sample (probability sample)*: A random sample is a sample taken from the population in such a way that every unit in the population as an equal chance (probability) of appearing in the sample. In research and scientific investigations the data collected by conducting an experiment repeatedly under identical conditions is considered as a random sample. Random samples are drawn from the population by using random number tables or by applying some random mechanism like chits.

(ii) *Non- random sample (non-probabilities sample):* Non random samples are collected by selection or judgement bases.   For example a sample of sixteen football players from a college for an intercollegiate competition are selected based on their skill in the game. i.e., by judgement. Random sample do not serve this purpose.

In research and scientific investigations the samples used shall be random samples only. The purpose of random samples is for applying various statistical methods for the analysis of data or result.

Objectives of Sampling

1.   Estimation of population parameters (mean, SD etc.) from the sample data.
2.   To test hypothesis about the population from which the sample or samples are drawn.

Variables

Variables are the measurable characteristics which can be numerically expressed in terms of some unit. These are quantities which are capable of being measured by quantitative methods directly.

For example, Height in inches, cm, weight in kg, pound, marks in examination etc.

Variables are of two types

(i)   **Discrete Variables (Discontinuous):** These are the quantities which can be measured in whole integral values. It does not take fractional value.

Example-Number of books, number of students in a college, number of defective items, family size etc.

(ii) **Continuous Variables:** These are quantities which can take any value which can be measured against a scale. It can take both integral and fractional values.

Example-Heights and weights

Yield of a chemical reaction

Globule size in an emulsion

Disintegration time of a tablet

Content of active ingredient in an ointment preparation

 ## Statistical Error

In statistical terminology the word 'error' is used in special sense. Error shows the extent to which the observed value of a quantity exceeds the true value.

Error = observed value − true value

Types of Error

Statistical error may be classified as:

(i) *Biased errors*: (which arise due to personal prejudices or bias.)

(ii) *Unbiased errors*: (which arise due to chance causes).

Array

The presentation of data in ascending order of magnitude is called array.

Method of Presentation of Statistical Data:

Statistical data are presented in three forms:

1. *Textual Presentation*

(i) Numerical data presented in a descriptive form are called textual presentation.

(ii) It is lengthy. Some words may repeat several times in the text.

(iii) It becomes difficult to grasp salient points in a textual presentation.

2. *Tabular Presentation*

(i) The logical and systematic presentation of numerical data in rows and columns designed to simplify the presentation and facilitate comparision is termed as tabulation.

(ii) Tabulating is thus a form of presenting quantitative data in condensed and concise form so that numerical figures are capable of easy and quick reception by the eyes.

(iii) It is more convenient than textual presentation.

3. *Graphical presentation*: The presentation of quantitative data by graphs and charts are termed as graphical presentation.

Tabulation: It may be defined as the logical and systematic presentation of the numerical data in rows and columns designed to simplify the presentation.

The advantages of tabulation are:

1. It enables the significance of the data readily understood and leaves a lasting impression than textual impression.

2. It facilitates quick comparison of statistical data shown between rows and columns.

3. Errors and omissions can be readily detected when data are tabulated.

Statistical Tables

Statistical tables are a systematic arrangement of quantitative data under appropriate heads in rows and columns. After the data have been collected they should be tabulated that is put in the form of a table so that the whole information can be had at a glance.

***Example* 1**

Table 4.2 Physical properties of telmisartan tablets prepared employing β-cd, primojel and tween 80.

Formulation Code	Hardness (Kg/sq.cm)	Friability (% wt Loss)	Disintegration Time (Sec)	Drug Content (%)
F 1	4.5	0.78	385	98.2
F a	4.0	0.92	345	99.6
F b	5.0	0.70	20	100.2
F ab	4.5	0.80	185	99.6
F c	5.0	0.85	390	98.4
F ac	4.5	0.75	190	99.2
F bc	4.5	0.80	40	98.9
F abc	5.0	0.90	55	98.8
F opt1	5.0	0.85	25	99.2

Parts of a Table

1. *Title*:
   (a) This is the brief description of the contents of the table.
   (b) The title should be clear and precise.
   (c) It should be at the top of the table.

2. *Stub*:
   (a) The extreme left part of the table where the description of the rows are shown is called stub.
   (b) It must be precise and clear.

3. *Caption and Box head*:
   (a) The upper part of the table which shows the description of columns and sub columns is called caption.
   (b) The whole of the upper part including caption units of measurement and column number if any is called box head.

4. *Body*:
   (a) It is the main part of the table except the title stub and captions.
   (b) It contains numerical information which are arranged in the table according to the description of the rows and columns given the stub and caption.

5. *Source and foot note*:
   (a) It is customary that source of the data from which information has been arrived should be given at the end of the table.
   (b) Foot note is the part below the body where the source of the data and any explanation are shown.

Essential Features of a Good Table

1. A table must have a title giving clear and precise idea about the contents of the table.

2. Units of the measurements adopted in a table must be shown clearly in the top of the column.

3. It is necessary that an investigator prepares a table well-proportioned in length and breadth.

4. For a compatible comparison, column of relevant figures must be kept as close as possible.

5. Distinction is preferred in columns and sub columns. It can be made by distinct ruling.

6. Totals of column may be shown in the bottom of the table. In case where row totals are useful, they should also be shown.

7. Table must contain necessary details.

8. Source of information must be disclosed at the end of the table.

9. Any ambiguous or confusing entry in the table should bear a special note at the end of the table for experiment.

10. The arrangement of the items in the table should have a logical sequence.

Example 2

Table 4.3 Fomulae of Telmisartan tablets prepared employing β-cd, primojel and tween 80 as per 2^3 factorial design.

Ingredient (mg/ tablet)	Formulation Code								
	F1	F a	F b	F ab	F c	F ac	F bc	F abc	F opt 1
Telmisartan	40	40	40	40	40	40	40	40	40
β-cylodextrin	40	240	40	240	40	240	40	240	140
Primojel	1.6	5.6	24	84	1.6	5.6	24	84	50.2
Tween 80	-	-	-	-	1.6	5.6	1.6	5.6	1.8
Talc	1.6	5.7	2	7.2	1.6	5.8	2.1	7.3	4.6
Magnesium stearate	1.6	5.7	2	7.2	1.6	5.8	2.1	7.3	4.6
Total Weight (mg)	84.8	297	108	378.4	86.4	302.8	109.8	384.2	241.2

F opt 1: Optimised Formulation to achieve NLT 85% Dissolution in 10 Minutes

Parameter and Statistic

Parameter is a statistical measure that describes the character of the population. Each population is characterized by two parameters, (i) average or mean and (ii) standard deviation. Population mean is denoted by μ and population standard deviation is denoted by σ .The population parameters are not known. The purpose of experiments is to give estimates for the unknown population parameters.

Statistic is an estimated quantity based on sampled data. For example the average estimated based on a sample of ten items is known as a statistic. The sample average is denoted by $\bar{x}$ and the sample standard deviation is denoted by s.d. The sample average $\bar{x}$ is taken as a measure or estimate of unknown population parameter μ. Sample average is not equal to the population average, μ. The sample average always includes some error and hence the sample average should always accompany a measure of error usually standard error (s.e). The other statistics include sample variance, s.d and correlation coefficient. A statistic characterises sample.

Concept of Probability: Probability is defined as the relative frequency of occurrence of an event. The probability is estimated as follows:

Probability = Number of favourable events / Total number of events

***Example* 1:** When a coin is tossed there are two possibilities, (i) getting Head and (ii) getting Tail. The total number of events in this case is 2 (two). If we are interested in getting head the favourable number of events is 1 (one). Then the probability of getting Head when a coin is tossed is ½ = 0.5. Similarly the probability of getting a Tail is also ½ = 0.5. The total probability is always equal to 1 (one).

***Example* 2:** A manufactured batch of Paracetmol syrup bottles contains 10 % defectives. What is the probability of getting a defective item and a non defective item from the batch?

In this case the number of defectives is given on percentage basis. The batch contains 10 % defectives which mean there are 10 defective items and 90 non defective items in every 100 items. The total number of events is 100. To calculate the probability of getting a defective item there are 10 favourable events. Hence the probability of getting a defective item from the batch is 10/ 100 = 0.1. Similarly to calculate the probability of getting a non defective item, the favourable number of events is 90 out of 100. Hence the probability of getting a non defective item from the batch is 90/ 100 = 0.9. The total probability is 0.1 + 0.9 = 1.0.

All the statistical methods are inherently based on the probability concepts only.

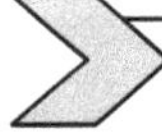# Presentation and Analysis of Data

Frequency Distribution

One of the major objectives of statistical methods is to describe appropriately the characteristics of mass of data. In any investigation, after collecting the necessary information, the next step should be to express it in some form which will permit conclusions to be drawn directly or by means of further calculations. It is not possible to detect relationships between various factors at issue from the unsorted mass of figures. The investigator must first consider the questions which he believes the material is capable of answering and then determine the form of presentation which brings out the true answers most clearly. The initial step must be to divide the observations into a relatively small series of groups, those in each group being considered alike in that characteristic for the purpose in hand. The tabulation or preparation of frequency tables thus constitutes an important step when the number of observations is large.

Frequency Distribution Tables and Graphs

Tabulation (or classification) is the process by which data of a long series of observations are systematically organized and recorded so as to enable analysis and interpretation. Classification is made on the basis of the size of individual observations. The range of observations is suitably divided into smaller divisions called class intervals and the number of observations falling in each of these intervals is counted and recorded.

The essential requirement of any such set of classes is that these classes should be selected in such a way that each one of the observations falls into one, and only one of these classes and that all the observations falls into any one of these classes. In setting up any table it is necessary to examine whether classification used is pertinent to the problem. A table must satisfy the following requirements:

1.   The table must be simple and clear-cut.
2.   The title of the table should explain in exact terms what the data represent.
3.   The figures in the body of the table must be arranged in a logical order for the point discussed in the text.
4.   When several points are to be emphasized making use of the same data, it may be preferable, whenever possible, to present the data in many small tables, one to illustrate each point.

A method of forming frequency distribution for a continuous variable is illustrated here for the 500 haemoglobin values. Glance through the data and pick out the lowest and the highest and divide the range of lowest and highest observations into 10 to 15 classes. For each observation, make a tally mark against the class in which it falls. For example, for the first observations, 11.6 make a tally mark against the class 11.5-12.4. After marking all observations, count the tallys and write the frequencies. This work sheet will look like a table. The table is called frequency distribution table (Table 4.4).

Table 4.4 Frequency distribution table of 500 haemoglobin values.

Class Interval	Tally Mark	Frequency
5.5-6.4	/	1
6.5-7.4	////////	8
7.5-8.4	///////////////////////////	31
8.5-9.4	//	44
9.5-10.4	/// ///////////	95

Table 4.4 *Contd….*

Class Interval	Tally Mark	Frequency
10.5-11.4	/// ///////////////////////////////////	125
11.5-12.4	/// ///////////////////////////////////	124
12.5-13.4	// /////////////////	49
13.5-14.4	////////	8
14.5-15.4	/////////	9
15.5-16.4	/////	6

 ## Diagrams

In addition to tables, graphs and diagrams are widely used in the presentation of statistical data either alone or in conjunction with the tables. They provide a visual method of examining quantitative data. They make a more lasting impression than detailed numbers and convey an idea forcefully. They help us to get a real grasp of the overall picture rather than the details.

In order that these graphs and diagram present ideas truthfully and emphasise correct ideas, they must be drawn following certain basic rules which are dependent partly on convention, partly on mathematical considerations, and partly on personal preferences.

Many types of graphs and diagrams are commonly used. Some important types are described below

Bar Diagram

A bar diagram is commonly used to provide a visual comparison of data of different groups. The frequency of each group is represented by a bar drawn for that group. Each bar must be of the same width. The height of the bar over a group represents the frequency of the corresponding group.

The Characteristics of the Bar Diagram are as follows:

(i)   It consists of a number of equally spaced rectangular areas with equal width and originates from a horizontal base line (X – axis).

(ii)   The length of the bar is proportional to the value it represents. It should be seen that the bars are neither too short nor too long.

(iii)   They are shaded or coloured suitably.

(iv)   The bars may be vertical or horizontal in a bar diagram. If the bars are placed horizontally, it is called as horizontal bar diagram, when bars placed vertically it is called a vertical bar diagram.

There are three types of bar diagrams:

1.   Simple bar diagram

2.   Multiple or grouped bar diagram

3.   Component or subdivided bar diagram.

1. Simple bar diagram

- It consist of a no.of equally spaced vertical bars of uniform width originating from a horizontal axis and is shaded.

- These bars are equally arranged according to relative MAGNITUDE OF bars.

- The length of the bar is determined by the value or the amount of variables.

- The limitation of simple bar diagram is that only one variable can be reperesented on it.

  Example: The heart beat rates of four mammals are given below. Represent it with the help of bar diagram.

2. Multiple or grouped bar diagram

- Multiple bar diagram reperesents more than one type of data at a time.

- In this case numerical values of major catagories are arranged in ascending or descending order so that catagories can be readily distinguished.

- Different shades or colour are used for each category.

- It is to be remember that the gap between the variable must be same.

3. Component or subdivided bar diagram

- Each bar in component bar diagram is sub divided into several component parts.

- A single bar represents the aggregate value whereas the component partsrepresents the component values of the aggregate value.

- It shows the relationship among the different parts and also between the different parts and the main bar.

- Different shades or colours are used to distinguish the various components.

Example: **Students of biological sciences of Serampore College during 2007 & 2008.**

Histogram: A histogram is a special kind of a diagram used to present a frequency distribution of a characteristic measured on a continuous scale. Rectangles are erected over class intervals to represent the frequencies of the class intervals. In a histogram, the area of each rectangle represents the frequency of the corresponding class interval.

The characteristics of the histogram are as follows:

The presentation of the distribution in Table 1 is shown in a histogram in the following figure.

- It consist of a set of rectangle drawn on a horizontal base line i.e., x-axis (abscissa) and frequency (i.e., no. of observations) is marked on the vertical line i.e., y-axis (Ordinate).

- The width of each rectangle extends over the class boundaries of the corresponding class along the horizontal axis.

- The area of each rectangle is proportional to the frequency in the respective class interval.

  The area each rectangle = width × height

$$= \text{Width of class} \times \text{frequency density}$$

$$= \text{width of class} \times \frac{\text{class frequency}}{\text{width of class}}$$

$$= \text{class frequency}$$

Types: There are two types of histograms :

(a) Histogram with equal class intervals

(b) Histograms with unequal class intervals

(a) Histograms with equal class intervals

  - Here the size of class intervals are drawn on x-axis with equal width and their respective frequancies on y-axis.

  - Class and its frequency taken together form a rectangle. The graph of rectangles is known as **histogram.**

  Example: Population of carp fishes in 100 ponds.

(b) Histogram with unequal class intervals

  - Here the sizes of class intervals are drawn on axis with unequal width and their respective frequencies on y-axis.

  - Some have less width and some have more width. So the

histogram is drawn on the basis of frequency density not on the basis of frequency.

- *Example*: Draw the histogram of the following frequency distribution with unequal class interval.

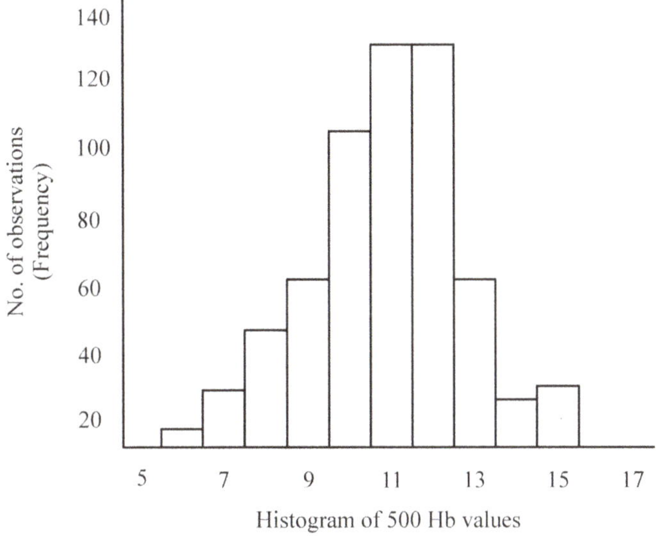

Figure 4.1

Frequency Polygon

A frequency polygon is a variation of histogram. Points of the tops of the rectangles in a histogram are joined by straight lines.

Frequency Curve

When the total frequency is large and when we adopt much narrower class intervals the frequency polygon will most often have a much smoother appearance. If the total frequency is increased indefinitely, the frequency polygon will approach a smooth curve. This limiting condition is known as the frequency curve.

A frequency curve generally describes the relationship between the frequency or relative frequency (probability) and magnitude of the variable. In theoretical studies we come across many such curves.

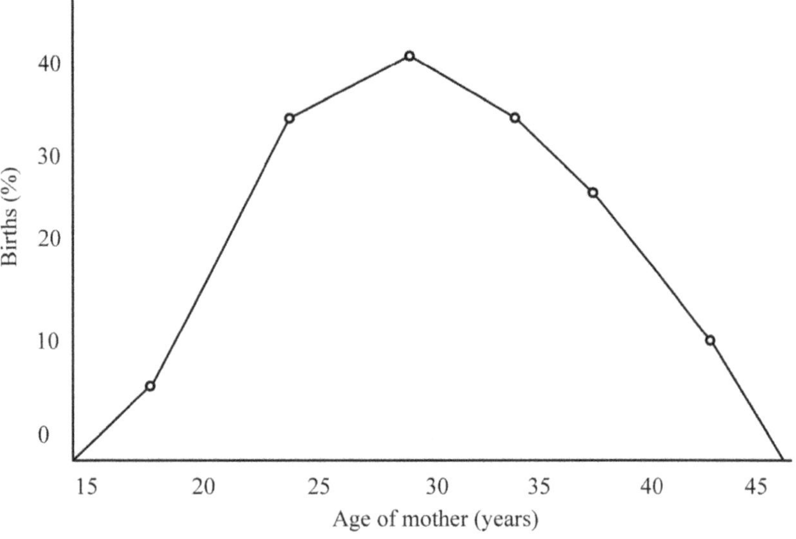

Figure 4.2

A model frequency distribution curve relating the age of mother and percentage birth is shown in figure above.

Characteristics of a Frequency Distribution

After forming a frequency distribution, the next step is the calculation of certain values which may be used as descriptive measures of the characteristics of that distribution. These values will enable comparisons to be made between one series of observations and another.

There are four major characteristics of a frequency distribution:

1. The measure of central tendency, location or position (average);
2. The degree of scatter of the observations around the measure of central tendency (variability, dispersion, spread);
3. The extent of symmetry in the shape of the distribution (skewness, asymmetry);
4. The flatness or peakedness of the distribution (kurtosis).

Of these four characteristics, the first two are most commonly employed in describing a set of data.

When more than one factor is observed, it may be necessary not only to describe each of these factors in terms of the four descriptive measures mentioned above but also in terms of the degree of association between various factors considered. The two most commonly used characteristics are (a) correlation, and (b) regression.

Pie Diagram or Chart: It is a circular graph whose area is subdivided into sectors by radii in such a way that the areas of the sectors are proportional to the percentage of components.

The Characteristics of the Pie Diagram are as follows:

(i)   The area of the circle represents the total value and the different sectors of the circle represent the different parts.

(ii) It is generally used for comparing the relation between the various components of a value and between components and the total value.

(iii) It gives comparative difference at a glance.

(iv) In pie chart or diagram, the data is expressed as percentage. Each component is expressed as percentage of the total value.

(v)   The name of the pie diagram is given to a circle diagram because in determining the areas of a circle we have to take into consideration a quantity known as 'pie' (π).

Chapter 5

Measures of Central Tendency and Dispersion

Introduction

LEARNING OBJECTIVES

- To know
- Definitions, properties and calculation of various measures of central tendency and dispersion.
- Mean, Median, Mode, Range, Mean deviation, Standard deviation, Variance, Coefficient of variance.

In a set of data the values of the extent of the observations are not equal, but we notice a general tendency of such observations to cluster around a particular level. In this situation it may be preferable to characterize each group of observations by such a level, which is called the central tendency of that group. This single value for each group of observations serves as a representative of that group. This level around which the observations tend to cluster may vary from group to group.

Several measures of central tendency can be calculated for a group of observations. The most widely used are mean, median and mode. Mean could be arithmetic, geometric or harmonic.

Arithmetic Mean

The arithmetic mean of a group is the simple arithmetic average of the observations. This is calculated by dividing the total of all the observations by the number of observations. In the case of grouped data (frequency distribution), arithmetic mean is calculated assuming that each observation in a class interval is equal to the midpoint of that class interval.

The following formulae are used to calculate the arithmetic mean. In an ungrouped data, if x represents the characters observed, and n the number of observations, then all the observations in the data can be denoted as x_1, x_2,......x_n.

The arithmetic mean is given by

$$m \text{ or } = \overline{x} = x_1 + x_2 + ... \ x_n \ / n = \Sigma x \ / n$$

For the grouped data (frequency distribution) the arithmetic mean is given by $\overline{x} = \Sigma fix_i \ / n$ where fi is the frequency, xi mid-point of the class interval and n the total number of observations.

Table 5.1 Calculation of Arithmetic Mean for Serum.

Albumin Levels (g%) of 24 Pre-School Children

2.90	3.75	3.66
3.57	3.45	3.76
3.73	3.71	3.43
3.55	3.84	3.69
3.72	3.30	3.77
3.88	3.62	3.43
2.98	3.76	3.68
3.61	3.38	3.76

The total of all these values i.e $\Sigma x_i = 85.93$

Total numbers of observations (n) = 24

$$\text{Mean} = \Sigma x_i \ / n = \frac{85.93}{24} = 3.58 \ g\%$$

Table 5.2 Calculations of Arithmetic Mean of Protein Intake of 400 Families (grouped data).

Protein intake/consumption unit/day (g)	No. of families	Mid-point of class interval	Multiply f and x
Class interval	Fi	xi	Fixi
15-25	30	20	600
25-35	40	30	1200
35-45	100	40	4000
45-55	110	50	5500
55-65	80	60	4800
65-75	30	70	2100
75-85	10	80	800
Total	400		19000

Arithmetic mean:

$$\Sigma f_i x_i / n = 30 \times 20 + 40 \times 30 + \ldots + 10 \times \frac{80}{400}$$

$$= \frac{19000}{400} = 47.50 \text{ g}$$

Median

The median is the magnitude of the observation which occupies the middle position when all the observations are arranged in order of the magnitudes. When there are an even number of observations in the group, the median is given by the average of the magnitudes of the middle pair of these observations, all the observations having been arranged in order of magnitude as before. In the case of grouped data, the median is calculated assuming that the observations in each class interval are uniformly distributed over that class interval. The formula for calculating the median for a grouped distribution is

$$\text{Median} = L + \left(\frac{n}{2} - F \right) \frac{C}{f}$$

Where L is the lower limit of the median class, n is the total number of observations, F the number of observations up to the median class, f the frequency (number of observations) in the median class and C the interval of median class. The class interval that contains the median is called the median class.

Table 5.3 Calculation of median for the data given in Table1.

Arranging all the 24 values in ascending order of magnitude, we get the following data

2.90	3.57	3.73
2.98	3.61	3.75
3.30	3.62	3.76
3.38	3.66	3.76
3.43	3.68	3.76
3.43	3.69	3.77
3.45	3.71	3.84
3.55	3.72	3.88

The 12th value is 3.66 and 13th is 3.68; median is the average of these two.

$$\text{Median} = 3.66 + \frac{3.68}{2} = 3.67 \text{ g\%}$$

Table 5.4 Calculation of median for the data of table 2.

Protein intake/consumption unit/day (g)	No. of families	Cumulative frequency
15-25	30	30
25-35	40	70
35-45	100	170
45-55	110	280
55-65	80	360
65-75	30	390
75-85	10	400
total	400	

Median class is 45-55 $n = 400$

Median $= L + (n/2 - F) \, C/f$

$= 45 + (200 - 170) \times 10/110 = 45 + 2.73 = 47.73$ g

Mode

Mode is the most frequently occurring value. In other words, the mode of a group of observations is the value around which the observations tend to be most heavily concentrated. For example, for the data given in table1 the observation 3.76 is most commonly occurring and hence the mode is 3.76. If we have a group of values such as 31, 33, 34, 36, 37, 39, 40 it is apparent that there is no mode. For a moderately asymmetric distribution, the mode can be calculated using the following empirical relationship:

Mode $= 3$ median $- 2$ mean

or using the formula

Mode $= L_M + d_1 C/ \, d_1 + d_2$

Where

$L_M =$ lower limit of modal class

$d_1 =$ frequency in modal class minus frequency in the preceding class

$d_2 =$ frequency in modal class minus frequency in the succeeding class

$C =$ class interval of modal class

For the data of Table 5.4

Mode $= 45 + 10 \times 10/10 + 30 = 47.5$

Selection of the Appropriate Measure of Central Tendency

When these measures of central tendency are considered as descriptive indices, we are faced with the problem of deciding which one of these to use to characterize a given set of data. The choice will depend upon the nature of the distribution of the observations and the concept of the central tendency which is desirable for our data. Here are a few guidelines:

1. If the data are symmetrically distributed or are approximately symmetrical, any one of these measures may be used because in a symmetrical distribution all these measures give identical values.

2. When the distribution of the observations is skewed, the arithmetic mean is usually not suitable. For a positively skewed series, the mean gives a higher value than the other two measures; and for a negatively skewed series, a lower value. It may be preferable to use the median or the mode which is typical.

3. When there are some observations which relatively deviate much more than the others in the series or when heterogeneity is suspected in the series, the median may be used, instead of the mean.

4. When subsequent computations involving a measure are necessary, the arithmetic mean has certain definite advantages.

5. When the concept of relative standing of the individual observations in the group is considered, the use of the median is desirable; whereas the concept of typical observation necessities the use of the mode.

Other Measures of Central Tendency

Geometric Mean

The geometric mean (GM) is usually more suitable as a measure of central tendency when the values change exponentially. If there are only two observations then the GM is square root of the product of 2 observations. If there are 3 observations, then it is the cube root of the product of the three observations. Thus, if there are n observations, the GM will be the nth root of the product of the n observations, viz.,

$$GM = \frac{\sqrt[n]{(x_1)(x_2)....(x_n)}}{}$$

when one takes the logarithm of both sides,

Log $GM = \Sigma(\log x)/n$ which shows that the logarithm of the geo metric mean is the arithmetic mean of the logarithm of the individual observations.

Example of geometric mean (GM) calculation: The number of bacteria ($\times 103$) observed in an experiment at hourly intervals are as follows

10, 25, 76, 148, 302

The geometric mean of these 5 values will be

$$5\sqrt{10 \times 25 \times 76 \times 148 \times 302} = 5\sqrt{33,968,960} = 61.07$$

The same result can also be obtained by taking the arithmetic mean of the logarithm of these 5 values and then determining the antilog as follows:

Value	log
10	1.0000
25	1.3979
76	1.8808
148	2.1703
302	2.4800

$$\text{Total} \div 5 = 8.9290 \div 5 = 1.7858$$

$$\text{GM} = \text{antilog}\,(1.7858) = 61.07$$

Harmonic Mean

The harmonic mean is used in situations where the reciprocals of the actual values seem more useful to determine the central tendency.

$$1/\,\text{HM} = \frac{\varepsilon\left(\dfrac{1}{x}\right)}{n}$$

In the other words, the reciprocal of the harmonic mean is the mean of the reciprocals of the individual observations.

Examples of Harmonic Mean (HM)

The distance (km) from an industrial establishment of 17 cases of chronic bronchitis are as follows:

0.8, 1.2, 3.2, 1.6, 0.7, 1.1, 2.7, 2.1, 1.3, 0.9, 1.3, 1.5, 1.1, 0.9, 1.8, 2.2, 2.4

Thus

$$1/\text{HM} = \text{Sum of } (1/x)/n$$

$$= \text{sum of } (1.25, 0.83, 0.31, 0.63, 1.43, 0.91, 0.37, 0.48, 0.77, 1.11,$$
$$0.77, 0.67, 0.91, 1.11, 0.56, 0.45, 0.42) \div n$$

$$= 12.98 \div 17$$

$$1/\text{HM} = 0.7635$$

Therefore, HM = 1.31km.

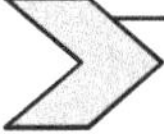 # Measures of Dispersion

The fact of that we need an average or a measure of central tendency shows that there is variation among the observations. Variation which is another characteristic of a group of observations has to be considered for describing the group more satisfactorily. A single figure for a group relating to its central tendency does not give any idea about the variability of the observations. In the case of the following haemoglobin values of two groups of children, they have the same central tendency (12.5), but the amount of spread of observations in each group differs very much

Group I: 12.1   12.2   12.8   12.9   12.3   12.4   12.7   12.6   12.5

Group II: 12.1   12.3   11.7   11.9   13.1   13.3   12.5   12.9   12.7

It is often necessary to give an idea of the quantity of such spread, or variability of the observations in a group while characterizing that group. For describing this, more than one measure is available. A measure of central tendency like mean along with a measure of dispersion form a fairly adequate presentation of a frequency distribution.

The measures of dispersion like range, mean deviation, standard deviation and coefficient of variation are also called measures of variation.

Range

The range of a group of observations is the interval between the smallest and the biggest observations. The value of the range is dependent only upon the two extreme observations in the group and does not consider the other observations. The occurrence of rare observations in the group greatly influence the value of the range and so it is not considered ideal as a measure of dispersion even though it is used in certain cases like for construction of control charts in statistical quality control.

Mean Deviation

The mean deviation is the arithmetic mean of the deviations of the observations from the arithmetic mean ignoring the sign of these deviations.

The mean deviation is based on all observations in this group. It is easy to grasp the meaning of the procedure involved in this measure. However, it is not widely used because of the availability of a more deviation for ungrouped data is

$$\text{Mean deviation} = \frac{\varepsilon|x - \bar{x}|}{n}$$

Where $|x - \bar{x}|$ indicates the difference between the value of the observation and the arithmetic mean ignoring the sign of the difference. For the grouped data (frequency distribution).

$$\text{Mean deviation} = \frac{\varepsilon f |x - \bar{x}|}{\varepsilon f}$$

Where x is the mid-point of the class interval and f is the frequency.

***Example* 1**: To calculate the mean deviation for the data given in table 5.5.

Table 5.5 Calculation of Mean Deviation.

Hb value (g%)	Deviation from arithmetic mean* (without sign)	Hb level (g%)	Deviation from arithmetic mean* (without sign)
11.8	0.4	11.7	0.5
11.4	0.8	12.7	0.5
10.4	1.8	12.2	0.0
11.6	0.6	11.6	0.6
10.8	1.4	12.6	0.4
12.2	0.0	13.3	1.1
12.9	0.7	12.9	0.7
12.3	0.1	13.2	1.0
10.8	1.4	14.2	2.0
12.0	0.2	13.5	1.3
10.5	1.7	13.0	0.8
11.2	1.0	13.8	1.6
12.4	0.2	12.2	0.0
	total	317.2	20.8

*mean = 12.2

Total of the deviations from arithmetic mean (without taking into account the sign)

$$= \frac{\varepsilon |x - \bar{x}|}{n} = 20.8/26 = 0.8g\%$$

***Example* 2:** To calculate the mean deviation for the example given in table 5.6

Table 5.6 Calculation of Mean Deviation.

Protein intake/consumption unit/day (grams)	No. of families	Mid-point of class interval	Deviation of mid-point from mean*	Absolute deviation fx
Class interval	F	x	x - $\overline{X}$	$f\left|x-\overline{x}\right|$
15-25	30	20	-27.5	825
25-35	40	30	-17.5	700
35-45	100	40	-7.5	750
45-55	110	50	+2.5	275
55-65	80	60	+12.5	1000
65-75	30	70	+22.5	675
75-85	10	80	+32.5	325
total	400			4550

*mean = 47.5

Mean deviation = 825 + 700+......+325/400 = 4550/400 = 11.375

Variance and Standard Deviation

The standard deviation is the square root of the average of the squared deviations of the observations from the arithmetic mean. The deviation from mean is considered without its sign in calculating the mean deviation, but in calculating the standard deviation, it is squared. The standard deviation is the most important measure of dispersion. The standard deviation together with the arithmetic mean can describe a frequency distribution uniquely. Standard deviation is a measure of error involved in the individual observation.

The standard deviation of the population is usually denoted by σ, and that of the samply by S.

$$\text{Standard deviation } (\sigma \text{ or } S) = \sqrt{\varepsilon\left|x-\overline{x}\right|^{2/n}}$$

For the grouped data (frequency distribution)

$$\text{Standard deviation } (\sigma \text{ or } S) = \sqrt{\varepsilon f\left|x-\overline{x}\right|^{2/ef}}$$

To get an unbiased estimate of population standard deviation from small samples, divide by n-1 instead of by n. this is denoted as 's'. Thus, the formula to compute standard deviation for small sample is

$$S = \sqrt{\varepsilon\left|x-\overline{x}\right|^{2/n-1}}$$

The square of standard deviation is called variance which can also be used as a measure of dispersion. Calculations of variance and standard deviation are shown in Table 5.7.

Table 5.7 Calculation of Standard Deviation for the Data of Table 1.

Serial no.	Haemoglobin values	Deviations From arithmetic mean 12.2	Square of deviation
1	11.8	-0.4	0.16
2	11.4	-0.8	0.64
3	10.4	-1.8	3.24
4	11.6	-0.6	0.36
5	10.8	-1.4	1.96
6	12.2	0.0	0.0
7	12.9	0.7	0.49
8	12.3	0.1	0.01
9	10.8	-1.4	1.96
10	12.0	-0.2	0.04
11	10.5	-1.7	2.89
12	11.2	-1.0	1.00
13	12.4	-0.2	0.04
14	11.7	-0.5	0.25
15	12.7	-0.5	0.25
16	12.2	0.0	0.0
17	11.6	-0.6	0.36
18	12.6	0.4	0.16
19	13.3	1.1	1.21
20	12.9	0.7	0.49
21	13.2	1.0	1.00
22	14.2	2.0	4.00
23	13.5	1.3	1.69
24	13.0	0.8	0.64
25	13.8	1.6	2.56
26	12.2	0.0	0
total	317.2	0	25.40

Arithmetic mean is 12.2

$$S = \sqrt{\varepsilon|x - \bar{x}|^{2/n-1}} = \sqrt{25.40/25}$$

$$s = 1.01g\,\%$$

Variance $= s^2 = 1.016$

Table 5.8 Calculation of standard Deviation for the grouped data.

Protein intake/consumption unit/day (grams)	No. of families	Mid-point of class interval	Deviation of mid-point from mean*	Squared deviation	Frequency × squared deviation
Class interval	f	X	$(x - \bar{x})$	$(x - \bar{x})^2$	$f(x - \bar{x})^2$
15-25	30	20	−27.5	756.25	22687.5
25-35	40	30	−17.5	306.25	12250.0
35-45	100	40	−7.5	56.25	5625.0
45-55	110	50	+2.5	6.25	687.5
55-65	80	60	+12.5	156.25	12500.0
65-75	30	70	+22.5	506.25	15187.5
75-85	10	80	+32.5	1056.25	10562.5
total	400				79500.0

*Arithmetic mean = 47.5

From this table, we get $\varepsilon f|x - \bar{x}|^2 = 79500.0$

$$\varepsilon f = 400$$

Therefore,

$$\text{Standard deviation} = S = \sqrt{79500/400} = 14.10G$$

Variance $= S^2 = 198.75$

Alternative Method of Calculating Standard Deviation

This method can be computationally easier using the same values and the alternative formula given

Table 5.9 Calculation of standard Deviation for the data.

S.No.	Haemoglobin values	Square of Haemoglobin values
1	11.8	139.24
2	11.4	129.96
3	10.4	108.16
4	11.6	134.56
5	10.8	116.54
6	12.2	148.84
7	12.9	166.41
8	12.3	151.29
9	10.8	116.64
10	12.0	144.00
11	10.5	110.25
12	11.2	125.44
13	12.4	153.76
14	11.7	136.89
15	12.7	161.29
16	12.2	148.84
17	11.6	134.56
18	12.6	158.76
19	13.3	176.89
20	12.9	166.41
21	13.2	174.24
22	14.2	201.64
23	13.5	182.25
24	13.0	169.00
25	13.8	190.44
26	12.2	148.84
Total	317.2	3895.24

Standard deviation, $s = \sqrt{\in x_2 - (\in x^2)/n/n-1}$

$$= \sqrt{3895324 - (317.2)^2 / 26 / 25}$$

$$= 1.01g \%$$

Table 5.10 Calculation of standard Deviation for the grouped data.

Protein intake/consumption unit/day (grams)	No. of families	Mid-point of class interval		Square of midpoint of class interval	Frequency × square
Class interval	f	X	f.x	x^2	$(f.x)^2$
15-25	30	20	600	400	12000
25-35	40	30	1200	900	36000
35-45	100	40	4000	1600	160000
45-55	110	50	5500	2500	275000
55-65	80	60	4800	3600	288000
65-75	30	70	2100	4900	147000
75-85	10	80	800	6400	64000
total	400		19000		982000

$$\text{Standard deviation} = \sqrt{1/400}\,(982000) - (19000)^2 / 400 = 14.10g\%$$

Coefficient of Variation

The coefficient of variation is the standard deviation expressed as a percentage of the arithmetic mean. When variability of groups of observations is dependent on the average size of the groups and / or when the observations are in different units of measurement, this measure of dispersion is adopted to facilitate comparison of the relative variability in groups or of different measurements.

Coefficient of variation (percent),

C. V. = standard deviation × 100/ mean

For the example given in table 1, the standard deviation, s = 1.01 and the arithmetic mean

$\bar{x}$ = 12.2, the coefficient of variation is (1.01/12.2) × 100 = 8.28%

Chapter 6

Tests of Significance

Tests of significance are statistical tests used to find out whether the observed result is significant or not. For example when a chemical reaction is carried out at two temperatures like $40^0\,C$ and $60^0\,C$, the average yield at the two temperatures was found to be $\overline{x}_{40}$ and $\overline{x}_{60}$ respectively. To evaluate the effect of temperature on the yield of chemical reaction we have to find out the difference in the two yields i.e., $(\overline{x}_{60} - \overline{x}_{40})$, which is taken as a measure of the effect of temperature on the yield of the chemical reaction. To know the effect of temperature on the yield of the chemical reaction we have to find out whether the observed result i.e. $(\overline{x}_{60} - \overline{x}_{40})$ is significant or not. The observed result $(\overline{x}_{60} - \overline{x}_{40})$ is due to two reasons, (i) chance factors and (ii) real factors.

Chance factors are minute differences or changes in the conditions of the experiment like temperature, pH, volume of reagents added etc. Individually these minute changes in the experimental conditions cannot be measured and controlled. Their occurrence in the experiments is also random. The effects of these chance factors are so minute that they cannot be measured. But these minute changes in the experimental conditions collectively

<table>
<tr><td>

LEARNING OBJECTIVES

♦ To know test of significance- general principle and practical procedure.

♦ Null hypothesis, alternative hypothesis, types of error

♦ Level of significance, P value

♦ Z-test of significance and its applications with worked out examples

♦ t-test of significance and its applications with worked out examples

♦ Analysis of paired data

♦ Testing the significance of correlation coefficient

♦ F-ratio test and its applications.

♦ Chi-square test of significance and its applications with worked out examples

</td></tr>
</table>

produce a measurable change in the response observed. When the observed result in the experimental investigations is due to these chance factors the result is said to be 'not significant'.

They observed result ($\bar{x}_{60} - \bar{x}_{40}$) is also due to real factors like a measurable differences in temperature, pH, concentration etc. In the said example of the chemical reaction the difference in the temperature studied i.e. 40^0 and 60^0 C is a real factor that may influence the yield of the reaction. When the observed difference in the yields i.e. ($\bar{x}_{60} - \bar{x}_{40}$) is due to a real factor it is said to be 'significant'.

Tests of significance are used to find out whether the observed result is due to chance factors or due to real factors. In other words to find out whether the observed result is significant or not significant.

Principle of Tests of Significance

In a typical test of significance a null hypothesis (H_0) is initially assumed. The null hypothesis is always an assumption of no difference in the two averages to be tested i.e $\bar{x}_{60} = \bar{x}_{40}$, which means that the factor temperature has no effect on the yield of chemical reaction. The yield of chemical reaction is influenced by chance factors only. Under the above null hypothesis (H_0) the probability of getting the observed difference, P ($\bar{x}_{60} - \bar{x}_{40}$) is estimated. If this probability is less than 0.05 i.e. (P < 0.05), the null hypothesis is rejected as events with such a low probability do not occur normally under the null hypothesis. Hence based on the observed data the null hypothesis is considered as wrong and is rejected. When the null hypothesis (H_0) is rejected an alternative hypothesis (H_1) is taken as $\bar{x}_{60} \neq \bar{x}_{40}$, which means that the yield at 40^0C and 60^0C is significantly different and the result is said to be significant.

If the probability of getting the observed difference i.e. P ($\bar{x}_{60} - \bar{x}_{40}$) > 0.05, the null hypothesis is accepted. Acceptance of null hypothesis indicates that the observed difference is due to chance factors only. The real factor temperature has no effect on the yield of the chemical reaction. When the H_0 is accepted the result is said to be 'not significant'. Acceptance of null hypothesis (H_0) also indicates that there is no sufficient evidence to establish the significance of the observed result.

The real problem in practice is that probabilities associated with the observed results cannot easily be estimated and hence a different procedure based on the above principle is followed for performing tests of significance.

Procedure of Tests of Significance or Steps Involved in Tests of Significance

Step 1: A null hypothesis (H_0) of 'No difference' is assumed. Under this null hypothesis there is no difference between the two averages to be tested and the observed result or difference is due to chance factors only. The real factor involved has no significant effect.

Test Criterion: The deviation of the observed result from the null hypothesis is measured with the help of a test criterion. The test criterion is a statistic which measures the deviation of the observed data from the null hypothesis. Statistics such as Z, t, $\mathcal{H}^2$, F-ratio are used as the test criteria in the tests of significance. Each test criterion is defined by a mathematical expression and has different application. For example, test criterion Z is defined as follows:

$$Z = \frac{|\mu - \bar{x}|}{\sigma / \sqrt{2}}$$ which follows a Normal distribution with (0,1) as parameters

The Z as defined above can be used to compare μ and $\bar{x}$ of large samples.

Step 2: An appropriate test criterion is selected and the value (magnitude) of the selected test criterion is estimated based on the experimental data.

Step 3: Decision making: If the test criterion value estimated based on the experimental data is greater than a critical test criterion value at a selected 'level of significance' the null hypothesis (H_0) is rejected as the observed data is deviated far away from the null hypothesis.

Step 4: Selecting Critical Test Criterion Value: For selecting critical test criterion value, the concepts of types of errors and level of significance are to be considered

Types of Errors and Level of Significance

In taking a decision on the acceptance or rejection of a null hypothesis (H_0) it is possible to commit the following two types of errors.

Type I Error: Rejecting a true null hypothesis is Type I error, which is considered as a serious error and has to be controlled in a test of significance.

Type II Error : Accepting a false null hypothesis is called Type II error.

Among the above the two types of errors type I error is considered as serious and probability of committing type I error is denoted as ' α '. The α multiplied by 100 i.e. $\alpha \times 100$ is known as level of significance (ls). In a test of significance the probability of committing type I error (α) is fixed at a suitable low level and care is taken to avoid type II error. The α is normally fixed at 0.05 and 0.01 levels. An ls of 5 % indicates that the α value is 0.05. In biological studies a 5% ls is normally used and in physical and chemical studies 1% ls is used. When a null hypothesis is rejected at 5% ls, the result is said to be 'significant' and when the null hypothesis is rejected at 1% ls, the result is said to be 'highly significant'.

At the chosen level of significance the critical test criterion value is taken from the corresponding distribution tables at appropriate degrees of freedom. The decision either to accept or reject the null hypothesis is taken by comparing the test criterion value based on the experimental data with the critical test criterion value obtained at the chosen level of significance. If the test criterion value based on the experimental data is greater than the critical test criterion value, the null hypothesis is rejected indicating that the observed result is significant or the difference in the two averages is significant and the factor involved is considered as having a significant effect.

If the test criterion value based on the experimental data is less than the critical test criterion value the null hypothesis is accepted as the observed result is in accordance of the null hypothesis. Accepting null hypothesis indicates that the observed result is 'not significant' or there is no sufficient evidence to prove that the observed result is significant.

The principle and procedure are same as described above for all the tests of significance like Z- test, t- test, F-ratio test, χ^2- test etc.

Z - test of Significance

Z – test of significance uses statistic Z as the test criterion. Standard normal variate (Z) follows normal distribution with μ and σ as parameters. Z is defined as follows

$$Z = \frac{|\bar{x} - \mu|}{\sigma / \sqrt{n}}$$ which follows normal distribution with 0 and 1 as mean and standard deviation.

Applications: Z as defined above is used for testing large samples.

1. To compare the average ($\bar{x}$) of a large sample with the population average (μ).
2. To test the significance of a single mean.
3. The test is employed to ascertain whether the population has a specified mean value based on some sample mean. For example to test whether the drug content of 100 liters of a suspension is 100 mg per 5 ml or not, based on a sample average of 30 samples, the Z-test can be used.
4. To compare proportions in two large samples.

5. To compare proportion of a large sample with a theoretical proportion as per null hypothesis.

   For testing proportions Z is defined as follows:

$$z = \frac{p-p}{\sqrt{\dfrac{p(1-p)}{N}}} \qquad N(0,1)$$

$$z = \frac{p_1 - p_2}{\sqrt{\dfrac{p(1-p)}{n_1} + \dfrac{p(1-p)}{n_2}}}$$

$$P = \frac{n_1 p_2 + n_2 p_2}{n_1 + n_2}$$

   Compare the calculated value of Z with the critical value of Z at a defined level of significance i.e 5 % ($Z_{0.05}$ = 1.96) or 1 % ($Z_{0.01}$ = 2.58). If the population standard deviation 'σ' is not known, its estimate given by sample standard deviation (s) is used.

6. The statistic Z is also useful to estimate confidence or fiducial limits.

   95 % confidence limits for μ are given by $\bar{x} \pm Z_{0.05}(\sigma/\sqrt{n})$ or $\bar{x} \pm 1.96\,(\sigma/\sqrt{n})$

   99 % confidence limits for μ are given by $\bar{x} \pm Z_{0.01}(\sigma/\sqrt{n})$ or $\bar{x} \pm 2.58\,(\sigma/\sqrt{n})$

Worked-out Examples

Example 1: A random sample of 400 subjects has an average haemoglobin content of 10 mg%. Can this be regarded as a sample from a large population with an average haemoglobin content of 10.2% and standard deviation (s.d) of 2.25 mg%. (Hint: 2 tail test)

Answer: Null hypothesis (H_o) : $\bar{x} = \mu$

   $n = 400$, $\bar{x} = 10$, $\mu = 10.2$, s.d $= 2.25$

   $Z = (10 - 10.2)/2.25/\sqrt{400}) = -\,0.02/0.1125 = -1.77$

Hence calculated Z is -1.77

Critical Z value ($Z_{0.05}$) $= 1.96$

As calculated Z value $< Z_{0.05}$ the null hypothesis (H_0) is accepted

Conclusion: The sample average and the population average of the same and hence the sample is regarded as a sample taken from the said population.

***Example* 2:** A random sample of 200 tins of a syrup gave an average weight of 4.95 kg with a standard deviation of 0.21. Do you accept the hypothesis of net weight of 5 kg per tin at 1% ls?

Answer: *Null hypothesis (H_o):* ($\overline{x} = \mu$), which means that the net weight of tin is 5 kg per tin

$$n = 200, \quad \overline{x} = 4.95, \quad \mu = 5.0, \quad s.d = 0.21$$

$$Z = (4.95 - 5) / 0.21 / \sqrt{200}) = -0.05 / 0.0148 = -3.378$$

Hence calculated Z is –3.378

Critical Z value ($Z_{0.01}$) = 2.58

As calculated Z value > $Z_{0.01}$ the null hypothesis (H_0) is rejected.

Conclusion: The net weight is not equal to 5 kg per tin

***Example* 3:** In a college of 500 students 280 are girls and the rest are boys. Can we assume that both girls and boys are in equal number in the college 1% ls?

Answer: *Null hypothesis (H_o):* Number of girls = Number of boys (or)

Proportion of girls = Proportion of boys

P = proportion of girls as per null hypothesis = 0.50

p = proportion of girls in sample = 280/ 500 = 0.56

$$Z(0.50 - 0.56)\sqrt{(0.50)(1 - 0.56) / 500} = -0.06 / 0.020 = -2.86$$

Calculated Z value = – 2.86

Critical Z value ($Z_{0.01}$) = 2.58

As calculated Z value > $Z_{0.01}$ the null hypothesis (H_0) is rejected.

Conclusion: The proportion of girls is not equal to proportion of boys.

Boys and girls are not in equal number.

***Example* 4:** A manufacturer claimed that at least 95% of the items he manufacture conformed to specifications. On examination of a sample of 200 items revealed that 18 were faulty. Test his claim at 5% ls? (Hint: One tail test)

Answer: *Null hypothesis (H_o):* $P \geq 95\%$ i.e. atleast 95% of items conform to specifications

$$n = 200$$

Number of items conforming to specifications $= 200 - 18 = 182$

p = proportion of items conforming $= (182 / 200) = 0.91$

$$Z = (0.91 - 0.95) / \sqrt{(0.95 \times 0.05)} / 200 = -0.04 / 0.015 = -2.67$$

Calculated Z value $= -2.67$

Critical Z value $(Z_{0.05}) = 1.645$ for one tail test

As calculated Z value $> Z_{0.05}$ the null hypothesis (H_0) is rejected.

Conclusion: The manufacturer's claim of atleast 95% of items conform to the specifications is false.

Example 5: A machine produced 20 defectives in a batch of 400. After overhauling it produced 10 defectives in a batch of 300. Has the machine improved? (Hint: One tail test)

Answer: $H_0 = p_1 = p_2$ i.e. the machine is not improved. The proportion of defectives before (p_1) is equal to the proportion of defectives after overhauling (p_2)

$$p_1 = 20 / 400 = 0.05; \; p_2 = 10/ 300 = 0.033$$

P = pooled proportion of defectives as per null hypothesis $= (20 + 10) / 400 + 300 = 3/70$

$$Z(0.05 - 0.033) / \sqrt{\{(3 / 70)(67 / 70)(1 / 400) + (3 / 70)(67 / 70)(1 / 300)\}}$$

$$= 0.017 / 0.0154 = 1.103$$

Calculated Z value $= 1.103$

Critical Z value $(Z_{0.01}) = 2.326$ for one tail test

As calculated Z value $< Z_{0.01}$ the null hypothesis (H_0) is accepted.

Conclusion: Machine has not improved after overhauling.

Example 6: In a group of 100 patients who received a drug, 75 were recovered from the disease. In a control group of 100 patients 65 were recovered. Test the efficacy of the drug at 5% le?

Answer: Null hypothesis H_0 : The drug is not effective. The proportion of recovery in drug treated group is equal to the proportion of recovery in control group.

Calculated Z value $= 1.538$

Critical Z value $(Z_{0.05}) = 1.96$

As calculated Z value $< Z_{0.05}$ the null hypothesis (H_0) is accepted.

Conclusion: The drug is not effective.

***Example* 7:** The mean IQ of a sample of 1600 children was 99. Is it likely that this was a random sample from a population with mean IQ 100 and standard deviation 15 at 5% ls.

***Solution*:** This case involves comparison of average ($\overline{x}$) of large sample with the population average (μ). For this purpose Z is defined as follows

$$Z = \frac{|\overline{x} - \mu|}{\dfrac{\sigma}{\sqrt{n}}}$$ which follows normal distribution with 0 and 1 as mean and standard deviation.

In the above case the sample average () = 99

Sample size, n = 1600

Population average μ = 100 with a standard deviation (σ) = 15

***Null hypothesis (H_0)*:** $\overline{x}$ = μ which means that the sample is taken from the said population

$$Z = \frac{|99 - 100|}{\dfrac{15}{\sqrt{1600}}} = \frac{|-1|}{\dfrac{15}{40}} = \frac{|-1|}{0.375} = 2.66$$

Therefore, Z value based on experimental data is 2.66

At 5% ls the critical Z value is 1.96. The Z estimated based on experimental data is greater than critical Z, hence null hypothesis is rejected.

***Conclusion*:** The sample is not taken from the said population, $\overline{x} \neq \mu$

***Example* 8:** A sample of 400 items is taken from a normal population whose mean is 4 and whose variance is also 4. If the sample mean is 4.45, can the sample mean be regarded as truly random sample.

***Solution*:** This case also involves comparison of sample average ($\overline{x}$) with population average (μ).

The population average μ = 4.0

Variance = 4

Hence standard variation (σ) = $\sqrt{4}$ = 2

Sample size (n) = 400

Sample average ($\overline{x}$) = 4.45

Null hypothesis (H_0) = $\overline{x} \neq \mu$ which mean that the sample is a random sample taken from said population.

For comparing ($\bar{x}$) and μ, Z is defined as follows

$$Z = \frac{|\bar{x} - \mu|}{\dfrac{\sigma}{\sqrt{n}}}$$

$$Z = \frac{|4.45 - 4|}{\dfrac{2}{\sqrt{400}}} = \frac{|0.45|}{\dfrac{2}{20}} = \frac{|0.45|}{0.1} = 4.5$$

Z-value based on experimental data is equal to 4.5.

Critical Z value at 5%ls is equal to 1.96.

As the Z value estimated based on experimental data is greater than critical Z value, the null hypothesis is rejected. Hence the random sample is not taken from the defined population, $\bar{X} \neq \mu$

***Example* 9:** A college conducts both day and night classes intended to be identical. A sample of 100 students from day classes yields examination results as under

$$\bar{X}_1 = 72.4 \text{ and } \sigma_1 = 14.8$$

A sample of 200 night students yields examination results as under

$$\bar{X}_2 = 73.9 \text{ and } \sigma_2 = 17.9$$

Are the means statistically equal at 5% ls.

***Solution*:** This case involves comparison of averages ($\bar{x}$) of two large samples. For this purpose

$$Z = \frac{|\bar{X}_1 - \bar{X}_2|}{\sqrt{\dfrac{\sigma_1^2}{n_1} + \dfrac{\sigma_2^2}{n_2}}}$$

Here in this case $\bar{X}_1 = 72.4$ and $\sigma_1 = 14.8$ (day time class), $n_1 = 100$

$\bar{X}_2 = 73.9$ and $\sigma_2 = 17.9$ (night time class), $n_2 = 200$

***Null Hypothesis*:** $\bar{X}_1 = \bar{X}_2$

$$Z = \frac{|72.4 - 73.9|}{\sqrt{\dfrac{14.8^2}{100} + \dfrac{17.92^2}{200}}} = \frac{|-1.5|}{\sqrt{2.19 + 1.602}} = \frac{|-1.5|}{\sqrt{3.792}} = \frac{|-1.5|}{1.947} = 0.77$$

Result: The Z value based on experimental data is 0.77. The critical Z value at 5% ls is equal to 1.96. As the Z value estimated based on experimental data is less than critical Z value, the null hypothesis is accepted. Hence the two averages $\bar{x}_1$ and $\bar{x}_2$ are equal. $\bar{x}_1 = \bar{x}_2$

***Example* 10:**   Two lots of manufactured tablets were tested for hardness. From each lot, 50 tablets were tested. The average and standard deviations were found to be 3.60 ± 1.8 and 4.48 ± 1.5 respectively for lot 1 and lot 2. Test at 5% ls whether the hardness is same in the two lots of the tablets manufactured.

Solution: This case involves comparison of averages of two large samples taken from two different lots of tablets manufactured.

Null Hypothesis: $\bar{x}_1 = \bar{x}_2$ i.e. hardness of tablets is same in the two lots. For this purpose Z is defined as

$$Z = \frac{|\bar{x}_1 = \bar{x}_2|}{\sqrt{\dfrac{\sigma_1^2}{n_1} + \dfrac{\sigma_2^2}{n_2}}} \sim N(0,1)$$

Here in this case for lot 1 $\bar{x}_1 = 3.60$ and $\sigma_1 = 1.8$, $n_1 = 50$ and for lot 2 $\bar{x}_2 = 4.48$ and $\sigma_2 = 1.5$, $n_2 = 50$.

$$Z = \frac{|3.60 - 4.48|}{\sqrt{\dfrac{1.8^2}{50} + \dfrac{1.5^2}{50}}} \frac{|-0.88|}{\sqrt{0.0648 + 0.045}} = \frac{|-0.88|}{\sqrt{0.1098}} = \frac{|-0.88|}{0.33} = 2.66$$

Result: The Z value based on experimental data is 2.66. The critical Z value at 5%ls is equal to 1.96. As the Z value estimated based on experimental data is greater than critical Z value at 5% ls, the null hypothesis is rejected. Hence the two averages $\bar{x}_1$ *and* $\bar{x}_2$ not equal. $\bar{x}_1 \neq \bar{x}_2$, which means that the hardness of the tablets in the two lots of manufactured tablets is significantly different.

t-test of Significance

t-test of significance is based on test criterion 't'. The statistic t follows t-distribution with (n-1) degrees of freedom. Degrees of freedom (v) refer to the number of free chances we have in reproducing the same statistic again. For several statistics like $\bar{x}$, the degrees of freedom (df) are (n-1) where 'n' is sample size or number of observations. The values of statistic 't' for various sample sizes at different levels of significance are available in standard t- distribution tables. The value of 't' varies with sample size 'n', where as the value of Z is independent of sample size.

The test criterion 't' is defined as follows

$$t = \frac{\overline{x} - \mu}{\sqrt{\dfrac{s^2}{n-1}}}$$

which follows t- distribution with (n-1) df.

Where $\quad S^2 = \dfrac{\sum x_i^2}{n} - \left(\dfrac{\sum x_i}{n}\right)^2$

Applications of t- test of Significance:

t-test of significance is used for testing small samples (n < 30).

To compare sample mean ($\overline{x}$) with population mean (μ).The statistic, 't' as defined above is used for this purpose.

To estimate fiducial or confidence limits based on sample data.

95 % confidence limits for μ are given by $\overline{x} \pm t_{0.05(v)} \left(S/\sqrt{n} - 1\right)$

99 % confidence limits for μ are given by $\overline{x} \pm t_{0.01(v)} \left(S/\sqrt{n} - 1\right)$

To compare the means of two small samples. For this purpose the 't' statistic is defined as follows

$$t = \frac{\overline{x}_1 - \overline{x}_2}{\sqrt{S^2\left((\dfrac{1}{n_1}) + (\dfrac{1}{n_2})\right)}}$$ which follows t- distribution with (n-2) df

Where, pooled variance, $S^2 = \dfrac{(n_1 S_1^2 + n_2 S_2^2)}{(n_1 + n_2 - 2)}$

To test the significance of mean difference of paired data:

The sample observations (x_1, x_2...........x_n) and (y_1, y_2, ... Y_n) are not completely independent, but pair wise related i.e. the pair of observations (x_1,y_1), (x_2,y_2).....(x_n,y_n) corresponds to 1, 2, 3........ n units respectively. The pair of observations (x_1,y_1) are obtained from the same experimental unit (subject or animal or machine etc.)

The null hypothesis (H_0) for comparison of paired data is $\overline{d} = \overline{0}$

where d is the average of the deviations of all the pairs of data.

The t-statistic for testing the paired data is as follows

$$t = \frac{\overline{d}}{\sqrt{s^2/n-1}}$$ Follows a t-distribution with (n-1) df

Where d = x-y: $\overline{d} = \Sigma d/n$

$$S^2 = \frac{\Sigma d^2}{n} - \left(\frac{\Sigma d}{n}\right)^2$$

For testing significance of correlation coefficient: To test the hypothesis that the correlation coefficient of the bi- variate normal population is zero, we can make use of the t-test. Let 'r' is the observed correlation coefficient, while p is the population correlation coefficient, the hypothesis to be tested is p = 0 i.e. population correlation coefficient is zero. Under this null hypothesis the test statistic is given by

$$t = \frac{r}{\sqrt{1-r^2/n-2}}$$

Which follows t- distribution with (n – 2) df

In order to decide whether a given correlation coefficient (r) indicates a linear relationship, we have to test the significance of 'r'

Worked-out Examples

***Example* 1:** A certain drug is packed into capsules by a machine. A random sample of 10 capsules is drawn and their contents are found to weigh in mg as follows

50, 49, 52, 44, 45, 48, 46, 45, 49, 45

Test if the average packaging can be taken as 50 mg.

Solution: This problem involves comparison of sample average ($\overline{x}$) with population mean (μ)

Null hypothesis (H₀) : $\overline{x} = \mu$ which means the average packaging is 50 mg

$\mu = 50$; $\overline{x} = 47.3$; $S^2 = 6.41$; $n = 10$;

$t = (43.7 - 50)/\sqrt{(6.41)/9} = -2.7/0.843 = -3.20$

Calculated t value = - 3.20

Critical t value $t_{0.05}(9) = 2.262$

As calculated t value > $t_{0.05(9)}$ the null hypothesis (H_0) is rejected.

Conclusion: The average packaging is not 50 mg.

Example 2: The average number of items produced by two machines per day are 200 and 250 with standard deviations of 20 and 25 respectively on the basis of records of 25 days production. Can you regard both the machines are equally efficient at 1% ls?

Solution: This problem involves comparison of two sample averages ($\bar{x}_1$ and $\bar{x}_2$)

Null hypothesis (H_0): Both the machines are equally efficient, ($\bar{x}_1 = \bar{x}_2$).

$$\bar{x}_1 = 200; \quad \bar{x}_2 = 250; \quad n_1 = n_2 = 25; \quad S_1 = 20; \quad S_2 = 25; \quad 1/n_1 = 1/25 = 0.04$$

Pooled variance, $\hat{S}^2 = 533.85;$

$$t = (200 - 250) / \sqrt{\{(5833.85)(0.04 \div 0.04)\}} = 50/6.535 = 7.65$$

Calculated t value = 7.65

Critical t value $t_{0.01}(48) = 2.58$

As calculated t value > $t_{0.01(48)}$ the null hypothesis (H_0) is rejected.

Conclusion: The machines are not equally efficient.

Example 3: A certain drug administered to each of 12 patients resulted in the following increase in blood pressure (B.P).

5, 2, 8, -1, 3, 0, -2, 1, 5, 0, 4, 6.

Can it be concluded that the drug will be associated by an increase in B. P.

(Hint: One tail test)

Solution: This problem involves testing paired data

Null hypothesis (H_0): Drug is not associated with an increase in B.P

$$\bar{d} = 0.8$$

S^2 of deviations = 12.36; n = 10;

$$t = 0.8 / \sqrt{(12.36/9)} = 0.8/1.171 = 0.683$$

Calculated t value = 0.683

Critical t value $t_{0.05}(9) = 1.833$ (for one tail test)

As calculated t value < $t_{0.05(9)}$ the null hypothesis (H_0) is accepted.

Conclusion: Drug is not associated with an increase in B.P

Example 4: A drug is given to 10 patients and the increase in their B.P. were recorded as 3, -6, -2, 4, -3, 4, 6, 0, 0, 2. Is it reasonable to believe that the drug has no effect on B.P.?

(Hint: Two tail test)

Solution: This problem involves testing paired data

Null hypothesis (H₀): Drug has no effect on B.P

$\overline{d} = 0.8$;

S^2 of deviations $= 12.36$; n $= 10$;

$t = 0.8\sqrt{(12.36/9)} = 0.8/1.171 = 0.683$

Calculated t value $= 0.683$

Critical t value $t_{0.05}(9) = 2.26$ (for two tail test)

As calculated t value$<t_{0.05(9)}$ the null hypothesis (H₀) is accepted.

Conclusion: The drug has no effect on B.P.

Example 5: In a nutritional study, 13 children were given a usual diet plus vitamin A and D tablets while the second comparable group of 12 children was taking the usual diet. After 12 months, the gain in weight in pounds was noted as given in the table. Can you say the vitamins A and D were responsible for this difference.

Group A : 5,3, 4, 3, 2, 6, 3, 2, 3, 6, 7, 5, 3

Group B : 1, 3, 2, 4, 2, 1, 3, 4, 3, 2, 2, 3, –

Solution:

Group A		Group B	
X_1	X_1^2	X_2	X_2^2
5	25	1	1
3	9	3	9
4	16	2	4
3	9	4	16
2	4	2	4
6	36	1	1
3	9	3	9
2	4	4	16
3	9	3	9
6	36	2	4
7	49	2	4
5	25	3	9
3	9	-	-

A	B
$$\bar{x}_1 = \dfrac{\sum x_1}{n} = \dfrac{52}{13} = 4$$	$$\bar{x}_2 = \dfrac{\sum x}{n} = \dfrac{30}{12} = 2.5$$
$$S_1^2 = \dfrac{\sum x_1^2}{n} - \left(\dfrac{\sum x_1}{n}\right)^2$$	$$S_2^2 = \dfrac{\sum x_2^2}{n} - \left(\dfrac{\sum x_2}{n}\right)^2$$
$$= \dfrac{249}{13} - (4)^2$$	$$= \dfrac{86}{12} - (2.5)^2$$
$$= 19.15 - 16$$	$$= 7.16 - 6.25$$
$$= 3.15$$	$$= 0.916$$

Pooled Variance

$$S^2 \frac{(n_1 S_1^2 + n_2 S_2^2)}{n_1 + n_2 - 2} = \frac{13 \times 3.15 + 12 \times 0.91}{13 + 12 - 2} = \frac{40.95 + 11}{23} = 2.25$$

$$t = \frac{\left|\bar{x}_1 - \bar{x}_2\right|}{S^2\left(\dfrac{1}{n_1} + \dfrac{1}{n_2}\right)}$$

$$t = \frac{\left|4 - 2\right|}{\sqrt{2.25}\left(\dfrac{1}{13} + \dfrac{1}{12}\right)}$$

$$t = \frac{\left|2\right|}{\sqrt{2.25}\left(0.076 + 0.083\right)}$$

$$= \frac{\left|2\right|}{\sqrt{0.35}} = \frac{\left|2\right|}{0.598} = 3.34$$

Result: The 't' value based on the experimental data is 3.34. Critical 't' value at 5% ls is 2.07. As the t value estimated based on experimental data is greater than critical t value at 5% ls, the null hypothesis is rejected. Hence the two averages $\bar{x}_1$ and $\bar{x}_2$ are significantly different. Hence the Vitamin A and D administration are responsible for this significant difference.

Example 6: The following data refers to the assay (mg/tablet) results of 5 tablet formulations before and after storage for 6 months in a stability testing programme.

Test the significance of the difference in assay content before and after storage.

Tablet	I	II	III	IV	V
Before storage	110	120	123	132	125
After storage	120	118	125	136	121

***Solution*:** This case is an example o,f analysis of paired data. The assay results before storage and after storage are taken as a pair. There are 5 such pairs. In each case the difference, d = X-Y, where X is before storage and Y is after storage is calculated. The significance of $\bar{d}$ is tested by t test of significance.

The null hypothesis (H₀) : $\bar{d} = 0$ which means there is no difference in the assay results before and after storage. For the analysis of paired data t is defined as follows.

$$t = \frac{\bar{d}}{\sqrt{\dfrac{S^2}{n-1}}} : t \text{ distribution with n-1 degrees of freedom}$$

$$\bar{d} = \frac{\sum d}{n}$$

$$S^2 = \frac{\sum d^2}{n} - \left(\frac{\sum d}{n}\right)^2$$

***Calculation*:**

Formulation	Assay Before Storage (X)	Assay After Storage (Y)	d (X-Y)	d²
I	110	120	−10	100
II	120	118	2	4
III	123	125	−2	4
IV	132	136	−4	16
V	125	121	4	16
			$\sum d = -10$	$\sum d^2 = 140$

$$\overline{d} = \frac{\sum d}{n} - \frac{10}{5} = -2$$

$$S^2 = \frac{140}{5} - (-2)^2$$

$$= 28 - 4$$

$$= 24$$

$$t = \frac{|-2|}{\sqrt{\dfrac{24}{4}}}$$

$$= \frac{|-2|}{\sqrt{6}} = \frac{|-2|}{2.44} = 0.819$$

The critical t value at 5% ls is equal to 2.78. As the t value based on experimental data is less than the critical t value at 5% ls, the null hypothesis is accepted. Hence $\overline{d} = 0$, which means the assay results before and after storage are the same.

***Example* 7**: In a sample of 42 pairs of x, y values the correlation coefficient is found to be 0.220. Does this indicate a significant correlation on the bases of 5% ls?

Solution: Null hypothesis (H_o) = 'r' is not significant

$$t \frac{0.220}{\sqrt{1 - (0.22)^2 / 40}}$$

$$= 0.220 / 0.154 = 1.43$$

The critical t- value, $t_{0.05(40)} = 2.021$

Since calculated 't' based on the observed data (1.43) is less than critical 't' value (2.021) the null hypothesis is accepted. Hence the correlation coefficient of 0.220 cannot be considered significant.

F- test of Significance

An F- test of significance is used to compare two variance quantities. For this purpose F- ratio is defined as follows

F = greater variance / lesser variance

The statistic 'F' follows F- distribution with (v_1, v_2) degrees of freedom. v_1 is the degrees of freedom of numerator and v_2 is the degrees of freedom of denominator.

***Example* 1:** A sample of $n_1 = 11$ items has a standard deviation (S_1) of 8.0. A second sample of $n_2 = 6$ items has a standard deviation (S_2) of 5.0. Do these sample standard deviations differ significantly at 5% ls?

***Answer*:**

$H_0 : S_1 = S_2$(no difference in the standard deviations)

$$n_1 = 11, S_1 = 8; v_1 = 11\text{-}1 = 10$$

Hence variance of the first sample is $8^2 = 64$

$$n_2 = 6, S_2 = 5; v_2 = 6 - 1 = 5$$

Hence variance of the second sample is $5^2 = 25$

F = greater variance/ lesser variance = 64/ 25 =2.56

Critical F value at 5% ls i.e. $F_{0.05(10,\ 5)} = 4.74$

As the calculated F value is less than the critical value the null hypothesis is accepted. Hence the difference in standard deviations is not significant.

χ^2 TEST (Chi – square Test of Significance)

Chi- square test is based on the statistic χ^2 as the test criterion. The statistic χ^2 is defined as

$$x^2 = \sum_{i=1}^{k}(0_i - e_i)^2 / e_i \text{ follows } \chi^2 \text{– distribution with (k- 1) df.}$$

Where o_i are the observed frequencies and e_i are the expected frequencies.

χ^2 measures the discrepancy between k number of observed frequencies (o_i) and their corresponding expected frequencies (e_i). The closure the agreement between expected and observed frequencies the smaller will be the value of χ^2. The statistic χ^2 follows a χ^2 distribution with (k-1) degrees of freedom.

Applications of Chi-square Test

1. To test the discrepancies between observed and expected frequencies.
2. To test goodness of fit. In regression analysis or curve fitting to know how well the curve fit the observed data.
3. To determine association between two or more attributes.
4. Chi square test is widely used to find whether or not any association between two or more attributes.
5. For example, we can find out whether there is an association between the colour of father's eye and son's eye.

In such cases the null hypothesis (H_0) is that there is 'no association' between the two attributes. If the chi square value calculated based on the observed value is less than the critical value or table value at a certain level of significance, the null hypothesis goes good and is accepted, otherwise the null hypothesis is rejected concluding that there is association between the two attributes.

Application of Chi-square for Testing Goodness of Fit

This involves testing whether observed results are consistent with the expected ones. The expected frequencies are worked out basing on a theory or null hypothesis.

Example **1:** According to Mandelian – inheritance theory in crossing two kinds of peas, four types of seeds A,B,C,D are expected to occur in the ratio of 9:3:3:1. In such an experiment, an experimenter obtains 102 seeds of type A, 30 of type B and 42 of type C and 15 of type D. Are these results consistent with the theory on the basis of 5% ls?

Solution: Null hypothesis (H_0) : The observed frequencies are in good agreement or consistent with the theory.

Calculation of expected frequencies (e_i):

Total seeds: $102 + 30 + 42 + 15 = 189$

According to theory expected type A seeds $= 9/\ 16 \times 189 = 106.3$ (observed 102)

expected type B seeds $= 3/\ 16 \times 189 = 35.4$ (observed 30)

expected type C seeds $= 3/\ 16 \times 189 = 35.4$ (observed 42)

expected type D seeds $= 1/\ 16 \times 189 = 11.8$ (observed 15)

Calculation of Chi-Square

Type	$(O_i - e_i)^2$	$(O_i - e_i)^2/\ e_i$
A	$(102 - 106.3)^2 = (4.3)^2 = 18.49$	$18.49/\ 106.3 = 0.173$
B	$(30 - 35.4)^2 = (-5.4)^2 = 29.16$	$29.16/\ 35.4 = 0.823$
C	$(42 - 35.4)^2 = (6.6)^2 = 43.56$	$43.56/\ 35.4 = 1.230$
D	$(15 - 11.8)^2 = (3.2)^2 = 10.24$	$1024/\ 11.8 = 0.867$ Total $= 3.093$

Calculated $\chi^2 = 3.093$

Critical $\chi^2_{\ 0.05}\ (\upsilon = 3) = 7.815$

Calculated χ^2 value is less than the critical χ^2. Hence H_0 is accepted i.e., the difference observed between o_i and e_i is only by chance.

Conclusion: There is good indication that the observed frequencies are in good agreement with expected frequencies as per Mandelian inheritance theory.

Application of χ^2 for testing the association or independency of attributes in contingency tables:

A contingency table is an arrangement in which a set of objects is classified according to two criteria of classification, one criterion being entered in rows and the other in columns. A 2×2 contingency table is widely employed. For a 2×2 contingency table the degree of freedom is taken as 1.

The null hypothesis (H_0) is no association between the attributes or the attributes are independent. This null hypothesis enables us to calculate expected frequencies.

Example 2: The following is the data on the number of men and women who became air sick on in airplane trip in rough weather. Do these data indicate women are less susceptible to air sickness than men at 5% ls?

	Susceptibility to airsickness		
	Number of cases of air sickness	Number of cases free of air sickness	Total
Men	24	30	54
Women	8	26	34
Total	32	56	88

Solution: The two criteria of classification in this case are (i) Sex (ii) Susceptibility to air sickness

Null hypothesis H_o : The two criteria of classification are independent.

Under this hypothesis we would expect the ratio of air sick men to all men to be the same as the ratio of the air sick women to all women and also the same as the ratio all air sick people to the total number of people of the plane.

Hence the expected frequencies are calculated basing on the ratio of all air sick people to the total number of people on the plane, which is equal to 32/88

Expected number of cases of airsick in men = 54×32/88 = 19.8

Expected number of cases free of airsick in men = 54-19 = 34.2

Expected number of cases of airsick in women = 34×32/88 = 12.2

Expected number of cases free of airsick in women = 34-12.2=21.8

Chi-Square Table

$o_i - e_i$	$(o_i - e_i)^2$	$(o_i - e_i)^2 / e_i$
24-19.8= 4.2	17.64	0.89
8-12.2 = -4.2	17.64	1.45
30-34.2= -4.2	17.64	0.52
26-21.8= 4.2	17.64	0.81
		Total = 3.67

Calculated $\chi^2 = 3.67$

Critical $\chi^2_{0.05}$ ($\upsilon =1$) = 3.84

Calculated χ^2 value is less than the critical χ^2. Hence H_0 is accepted

The two criteria of classification i.e. sex and susceptibility to air sickness are independent. Hence the data given do not indicate that women are less susceptible to air sickness than men.

Example 3: In tossing a coin 40 times, 25 heads and 15 tails are obtained. Is the coin biased or weighted in some way?

Solution: Null hypothesis (H$_0$) : The coin is unbiased

Expected frequency = Probability × number

Expected frequency of Head = ½× 40 = 20

Expected frequency of Tail = ½ × 40 = 20

Chi-Square Table

	o_i	e_i	$(o_i - e_i)^2$	$(o_i - e_i)^2 / e_i$
Head	25	20	$(25 - 20)^2 = 25$	25/20 = 1.25
Tail	15	20	$(15 - 20)^2 = 25$	25/20 = 1.25
				Total= 2.5

Calculated $\chi^2 = 2.5$

Critical $\chi^2_{0.05}$ ($\upsilon =1$) = 3.84

Calculated χ^2 value is less than the critical χ^2. Hence H_0 is accepted

Conclusion: The coin is unbiased

Example 4: The following are the survival rates of drug treated and control pigs with swine dysentery.

Treatment	Effectiveness		Totals
	Survival	**Dead**	
Drug treated	25	14	39
Control	21	22	43
Totals	46	36	82

Test the hypothesis that there is no difference in the survival rates of drug treated and control pigs. (or test whether the drug is effective)

Solution: *Null hypothesis (H_0)* : No difference in the survival rates of drug treated and control groups i.e drug is not effective.

Expected number of survivals in drug treated group = 39×46/ 82 = 21.87

Expected number of survivals in control group = 43×46/82 = 24.12

Expected number of dead in drug treated group = 39 × 36/82 = 17.12

Expected number of dead in control group = 43×36/82 = 18.87

Chi-Square Table

O_i	e_i	$(O_i- e_i)$	$(O_i- e_i)^2$	$(O_i- e_i)^2/ e_i$
25	21.87	3.13	9.796	0.447
21	24.12	-3.12	9.734	0.403
14	17.12	-3.12	9.734	0.568
22	18.87	3.13	9.796	0.519
				Total =1.937

Calculated $\chi^2 = 1.937$

Critical $\chi^2_{0.05}$ ($\upsilon =1$) = 3.84

Calculated χ^2 value is less than the critical χ^2. Hence H_0 is accepted

Conclusion: No difference in the survival rates of the drug treated and control groups. The drug is not effective.

Chapter 7

Analysis of Variance

Analysis of variance (ANOVA) is a statistical test of significance to compare several averages, more than two, to find out the significant differences among them.

Principle: In the ANOVA the total variance present in the data is analyzed in two different ways and the total variance is divided into components that arises due to (i) chance factors and (ii) real factors like temperature, binder, pH etc..

The two variance quantities are then compared by an F – ratio test. A significant F-ratio indicates that the treatment effects are significant (or) the averages are differing significantly among themselves.

In pharmacy research several times we need to compare more than two averages to find out the effects of various factors on the phenomena being studied. Some examples are as follows:

1. To study the effect of four binders on the dissolution rate of tablets we need to compare four averages corresponding to four binders.

2. To study the effect of three pHs on the stability of a drug solution we need to compare three averages corresponding to three pHs.

3. To study the effect of four temperatures on the yield of a chemical reaction we need to compare four averages corresponding to four temperatures.

4. To evaluate the dissolution rate of four brands of paracetamol tablets we need to compare four averages corresponding to four brands.

5. For bioequivalence testing of four drug products we need to compare four averages of bioavailability parameters such as C_{max} or AUC

 # ANOVA One Way Classification

In the ANOVA one way classification the data is divided in one direction based on one factor of variability. For example, in an experiment where the effect of four temperatures say 40^0, 50^0, 60^0, 70^0C on the yield of a chemical reaction is studied with a replication of $n = 4$, there are $4 \times 4 = 16$ observations representing the yields obtained in the 16 experiments. The total variation present in the 16 observations is due to (i) chance factors and (ii) real factor (treatment - temperature).

The total variance here is divided into two components, one that arises due to chance factors and other that arises due to treatment (temperature). For, this purpose the data is classified in one direction basing on the temperature i.e., all the observations obtained at 40^0C are considered as one block and those obtained at 50^0 C are considered as another block etc as shown below

Temp (^{0}C)	Yields in trials (n= 4)			
	1	2	3	4
40	$Y_{40,1}$	$Y_{40,2}$	$Y_{40,3}$	$Y_{40,4}$
50	$Y_{50,1}$	$Y_{50,2}$	$Y_{50,3}$	$Y_{50,4}$
60	$Y_{60,1}$	$Y_{60,2}$	$Y_{60,3}$	$Y_{60,4}$
70	$Y_{70,1}$	$Y_{70,2}$	$Y_{70,3}$	$Y_{70,4}$

The variance present in each block (group) is calculated from its average and then pooled to obtain the chance variance that arises due to chance factors.

Variance present in the data is also calculated in a different way based on the averages corresponding to each temperature. This variance includes the chance variance as well as the variance due to treatments (temperature). If the variance calculated based on the averages is equal to chance variance the treatment effects are considered not significant. If the variance calculated based on averages is significantly greater than the chance variance, the treatment effects are considered significant. The two variance quantities are then compared by an F–ratio test. A significant F-ratio indicates that the treatment (temperature) effects are significant at a particular level of significance.

The ANOVA one way classification is performed as per following table

Source of Variation	d.f	S.S	Mean Sum of Squares	F - ratio
Total	$N-1$	$\Sigma xij^2 / T^2/N$	TSS /df	
Treatment	$k-1$	$\Sigma T_i^2 / r - T^2/N$	Tr SS/ df	Mean Tr SS / Mean Er SS
Error (or) chance	$N-k$	TSS–Tr SS	Er SS/ df	

N = total number of observations; k = number of treatments; r = number of replications;

T = total of all observations; T_i = total of all observations of i th treatment; Σxij^2 = sum of squares of all observations

Examples:

1. To study the effect of temperature on the yield of a chemical reaction.
2. To study the effect of binder on the dissolution rate of tablets.
3. To compare four fertilizers on the yield of a crop

In the above examples, there is only one factor, the effect of which is studied. Hence the total variability in the data in each case is due to two factors, (i) chance factors and (ii) real factor involved. In such studies the results obtained are analyzed as per ANOVA one way classification.

ANOVA One Way – Worked-out Examples

***Example* 1:** An experiment with five samples of four replicates each. Are there significant differences in the means of five samples?

Solution: *Null hypothesis H_0:* There is no difference in the sample means; all samples are from the same population. No difference in the five means.

Sample (Treatment)

Replicates	A	B	C	D	E
1	10	12	9	14	10
2	4	10	4	9	6
3	6	13	4	10	8
4	4	7	5	11	4
Total	24	42	22	44	28
Mean	6.0	10.5	5.5	11.0	7.0

Total of all observations, T = 160

$$T^2/kn = (160)2/20 = 1280$$

$\Sigma\, xij^2$ = Sum of squares of all observations= 1482

$$TSS = 1482 - 1280 = 202$$

$$TrSS = 106\;;\; ESS = 202 - 106 = 96$$

ANOVA Table

Source of Variation	Sum of Squares	df	Mean Sum of Squares (MSS)	F - ratio
Total	202	19		
Between treatments	106	4	26.5	4.14
Within samples (Error)	96	15	6.4	

Calculated F-ratio = 4.14

Critical F-ratio = $F_{0.95}$ ($v_1 = 4$; $v_2 = 15$) = 3.06

Result: Calculated F-ratio > Critical F-ratio

Hence H_0 is rejected. Hence the five means are differening significantly.

The difference in means is significant

Extension of ANOVA

Which pairs of samples have significantly different means?

For knowing which pairs of samples have significantly different means we have to consider the concept of Least Significance Difference (LSD) orCritical Difference (CD)

Least Significance Difference (LSD) or Critical Difference (CD) is defined as follows

$$LSD = t_{0.05}\sqrt{2\,\frac{EMSS}{n}}$$

Where $t_{0.05}$ is 't' value corresponding to sample size or number of replicates at 5% ls obtained from t – distribution table

$$LSD = 2.131\sqrt{\frac{2x6.4}{4}}$$

$$= 3.812$$

For significant difference of means of two samples the difference in the two sample means should be > LSD

Means: 6.0(A) 10.5(B) 5.5(C) 11.0(D) 7.0(E)

The sample means of A and B differ significantly, since the difference 4.5 exceeds 3.812

Sample means differ significantly for the following pairs of samples: AB, AD, BC, CD, DE.

***Example* 2:** Three products of paracetmol tablets were evaluated for dissolution rate (mg/min) each four times. The dissolution rate values are given in the following table. Test the significance of product differences with regard to dissolution rate at 5%ls.

***Solution*:** This is an example of ANOVA one way classification. The one factor involved is product. Hence the total variance in the data is divided in to two components namely variance due to products and variance due to chance factors.

Null Hypothesis: H_0: The product differences are not significant

Replication	Products(Treatments)			
	A	B	C	
1	8	7	2	
2	7	5	3	
3	6	5	3	
4	7	6	4	
Ti	28	23	12	63
Ti^2/r	784/4 = 196	529/4 = 132.25	144/4 = 36	$\Sigma Ti^2/r =$ 364.25

$$Tss = \Sigma X_{ij}^2 - T^2/N$$

T = Total of all values

N = total number of values = 9

$$T^2/N = \frac{(63)^2}{12} = 330.75$$

ΣX_{ij}^2 = sum of squares of all values

	Products			
	A	B	C	
1	64	49	4	
2	49	25	9	
3	36	25	9	
4	49	36	16	
X_{ij}^2	198	135	38	$\Sigma X_{ij}^2 = 371$

$$T_{SS} = \Sigma X_{ij}^2 - T^2/N$$

$$= 371 - 330.75$$

$$= 40.25$$

$$T_{rss} = Trss_{(product)} = \frac{\Sigma T_i^2}{r} - \frac{T^2}{N}$$

$$= 364.25 - 330.75$$

$$= 33.5$$

$$E_{SS} = T_{SS} - T_{rss}$$

$$= 40.25 - 33.5$$

$$= 6.75$$

ANOVA Table

Source of Variation	df	SS	Mss = (SS/df)	Fratio	$F_{0.05\ (x,x)}$
Total	11	40.25	3.70	-	
Treatment (Product)	2	33.5	16.75	16.75/0.75 = 22.33	$F_{0.05(2,9)}$ = 4.26
Error	9	6.75	0.75		

Results: The calculated F ratio based on experimental data is greater than the critical F ratio at 5%ls, hence H_0 is rejected.

Hence product differences are significant.

***Example* 3:** The following data refers to percent reduction in blood glucose levels following the administration of three plant extracts to 4 rabbits each. Test the significance of the effects of plant extracts.

Plant Extract (Treatment)	Rabbit (Replication)			
	1	2	3	4
A	13	22	18	20
B	16	24	17	19
C	05	04	01	04

***Solution*:** This is an example of ANOVA one way classification. The one factor involved in this case plant exacts. The total variance in the data is due to two factors namely plant exact and chance factors. Hence the total variance in the data is divided into two components that arise due to plant extracts and chance factors.

Null hypothesis H_0: Effects of plant extracts are not significant

Plant Extract	Rabbit				T_i	T_i^2/r
	1	2	3	4		
A	13	22	18	20	73	1332.25
B	16	24	17	19	76	1444
C	05	04	01	04	14	49
					$\Sigma T_i = 163$	$\Sigma T_i^2/4 = 2825.25$

$$T_{ss} = \Sigma X_{ij}^2 - T^2/N$$

T = Total of all values

N = total number of values = 12

$$T^2/N = \frac{(163)^2}{12} = 2214.08$$

ΣX_{ij}^2 = sum of squares of all values

Plant Extract	Rabbit				X_{ij}^2
	1	2	3	4	
A	169	484	224	400	1377
B	256	576	289	361	1482
C	25	16	01	16	58
					$\Sigma X_{ij}^2 = 2917$

$$T_{ss} = \Sigma X_{ij}^2 - T^2/N$$
$$= 2917 - 2214.08$$
$$= 702.92$$

$$T_{rss} \text{ (plant extract)} = \frac{\Sigma T_i^2}{r} - \frac{T^2}{N}$$
$$= 2825.25 - 2214.08$$
$$= 611.17$$

$$ESS = TSS - Trss$$
$$= 702.92 - 611.17$$
$$= 91.75$$

ANOVA Table

Source of Variation	df	SS	Mss = (SS/df)	F ratio	$F_{0.05\,(v1,v2)}$
Total	11	702.92	63.9	-	
Treatment	2	611.17	305.83	305.83/10.19 = 30.01	$F_{0.05(2,9)} = 4.26$
Error	9	91.75	10.19		

Result: As the F-ratio based on experimental data is greater than critical F ratio, the null hypothesis is rejected. Hence the effects of plant extracts are significant.

***Example* 4:** The following data are the disintegration times in min of 4 tablet formulations each replicated 5 times. Test the significance of the difference in disintegration times at 1% ls.

Replicates	Formulation (Treatments)			
	F1	F2	F3	F4
1	1	2	3	5
2	2	3	3	4
3	1	3	4	5
4	0	4	3	4
5	1	3	4	4

Solution: This is an example of ANOVA one way classification. The one factor involved is formulations. The total variance in the data is due to two factors namely formulation and chance factors. The total variance is divided into two components, (i) variance due to formulation (ii) variance due to chance factors.

Null Hypothesis H_0: The formulation differences are not significant

Replicates	Formulation				
	F1	F2	F3	F4	
1	1	2	3	5	
2	2	3	3	4	
3	1	3	4	5	
4	0	4	3	4	
5	1	3	4	4	
	5	15	17	22	$\Sigma T_i = 59$
T_i^2/r	5	45	57.8	96.8	$\Sigma T_i^2/r = 204.6$

$$Tss = \Sigma X_{ij}^2 - T^2/N$$

T = Total of all values

N = total number of values = 20

$$T^2/N = \frac{(59)^2}{20} = 174.05$$

ΣX_{ij}^2 = sum of squares of all values

Replicates	Formulation				
	F1	**F2**	**F3**	**F4**	
1	1	4	9	25	
2	4	9	9	16	
3	1	9	16	25	
4	0	16	9	16	
5	1	9	16	16	
	7	47	59	98	$\Sigma X_{ij}^2 = 211$

$$T_{SS} = \Sigma X_{ij}^2 - T^2 / N$$
$$= 211 - 174.05$$
$$= 36.95$$

$$T_{rss \, (Formulations)} = \frac{\Sigma T_i^2}{r} - \frac{T^2}{N}$$
$$= 204.6 - 174.05$$
$$= 30.55$$

$$E_{SS} = T_{SS} - T_{rss}$$
$$= 36.95 - 30.55$$
$$= 6.4$$

ANOVA Table

Source of Variation	df	SS	Mss = (SS/df)	Fratio	$F_{0.05 \, (v1,v2)}$
Total	19	36.95	1.94	-	
Treatment	3	30.55	10.18	10.18/0.4 = 25.45	$F_{0.05(3,16)}$ = 4.26 $F_{0.01 \, (3,16)}$ = 5.29
Error	16	6.4	0.4		

Result: As the F-ratio based on experimental data is greater than critical F-ratio, the null hypothesis is rejected. Hence the formulation differences in disintegration time are significant.

ANOVA Two Way Classification

In the ANOVA two way classification the data are divided in two directions based on two different factors like temperature and p_H. All the observations obtained at a particular treatment(temperature) are considered as one group based on temperature, and all the observations obtained at a particular pH are considered as one group based on pH (block). The total variation present in the data is due to three factors,(i) chance factors (ii) treatment factor temperature and (iii) block factor pH. The total variance is divided into three components (i) variance due to chance factors (ii) variance due to treatment (temperature) and (iii) variance due to block (pH).

The variance estimates due to (i) temperature and (ii) pH are separately compared with the chance variance by an F – ratio test in each case. Significant F - ratios indicate that the effects of (i) temperature and (ii) pH are significant.

The ANOVA two way classification is performed in the following table:

Source of Variation	d f	S.S	Mean Sum of Squares	F - ratio
Total	$N-1$	$\Sigma xij^2 / T^2/N$	TSS /df	__________
Treatment (temperature)	$k-1$	$\Sigma T_{ik}^2 /r - T^2/N$	Tr SS/ df	Mean Tr SS / Mean Er SS
Block (pH)	$r-1$	$\Sigma T_{ir}^2 /k - T^2/N$	B SS / df	Mean B SS / Mean Er SS
Error (or) chance	$N-k-r+1$	TSS–(Tr SS + BSS)	Er SS/ df	

Examples

1. To study the effect of temperature and pH on the yield of a chemical reaction.
2. To study the effect of binder and disintegrant on the dissolution rate of tablets
3. To study the effect of pH and aeration rate on the yield of an antibiotic production.

In all the above cases the effect of two factors are studied simultaneously in each case.

The total variance in each case is due to

(i)   Chance factors

(ii)   Treatment factor (like temperature) and

(iii)   Block factor (like pH)

The total variance in the data is divided into three components as follows

    (i)   Variance due to Chance factors

    (ii)   Variance due to Treatment factor (temperature) and

    (iii)   Variance due to Block factor pH

The variance estimates due to treatment and due to block are separately compared with the chance variance by an F – ratio test. Significant F- ratios indicate that the effects of the two factors studied (temperature and pH) are significant and the averages due to factor 1(temperature) and due to factor 2 (pH) are differing significantly among themselves. In the ANOVA two way classification two conclusions can be taken related to the factors involved

ANOVA Two Way Classification

Worked out Example

***Example* 1:** Three drug products A, B, C were evaluated for bioavailability in 4 subjects each. The measurements of AUC with various products are given in the following table. Test the significance of product differences and subject differences at 5% ls.

Product	Subject			
	1	2	3	4
A	26	28	27	31
B	18	21	19	22
C	16	19	18	19

***Solution*:** It is an example of ANOVA with two way classification. The two factors involved are (i) drug product and (ii) subjects. Here the total variance in the data is due to products, subjects and chance factors. Hence the total variance is divided into the following three components,

    (i) Variance due to drug products (treatment)

    (ii) Variance due to subjects (blocks)

    (iii) Variance due to chance factor (chance variance)

When the measurements are large as in this case, the magnitude of the measurements is reduced by subtracting a constant value from all the measurements. When a constant value is subtracted from all the values in the data, the variance in the data does not change. In the present case 20 is subtracted from each measurement in the data and the ANOVA is performed on the resulting data.

Null Hypothesis:

$H_{0,1}$: Product differences are not significant.

$H_{0,2}$: Subject differences are not significant.

Product	Subject (Blocks)				T_i	T_i^2/r
	1	2	3	4		
A	6	8	7	11	32	256
B	-2	1	-1	2	0	0
C	-4	-1	-2	-1	-8	16
T_i	0	8	4	12	$\Sigma\,T_i = 24$	$\Sigma\,T_i^2/r = 272$
T_i^2/K	0	21.3	5.3	48	$\Sigma\,T_i^2/K = 74.6$	

$$Tss = \Sigma X_{ij}^2 - T^2/N$$

T = Total of all values

N = total number of values = 12

$$T^2/N = \frac{(24)^2}{12} = 48$$

ΣX_{ij}^2 = sum of squares of all values

Product	Subject				
	1	2	3	4	
A	36	64	49	121	
B	4	1	1	4	
C	16	1	4	1	
	56	66	54	126	$\Sigma X_{ij}^2 = 302$

$$Tss = \Sigma X_{ij}^2 - \frac{T^2}{N} = 302 - 48$$

$$= 254$$

$$Trss_{(product)} = \frac{\Sigma T_i^2}{r} - \frac{T^2}{N}$$

$$= 272 - 48 = 224$$

$$Blss_{(subjects)} = \frac{\Sigma T_i^2}{K} - \frac{T^2}{N}$$

$$= 74.6 - 48$$

$$= 26.6$$

$$Ess = T_{SS} - (T_{rss} + Bl_{ss})$$

$$= 254 - (224 + 26.6) = 3.4$$

ANOVA Table

Source of Variation	df	SS	Mss = (SS/df)	Fratio	$F_{0.05\ (x,x)}$
Total	11	254	23.0	-	
Treatment (Products)	2	224	112	112/0.56 = 200	$F_{0.05(2,6)}$ = 5.14
Block (subject)	3	26.6	8.86	8.86/0.56 = 15.82	$F_{0.05\ (3,6)}$ = 4.76
error	6	3.4	0.56		

Results:

1. As the F-ratio based on experimental data is greater than critical F-ratio, the null hypothesis $H_{0,1}$ is rejected. Hence the product differences are significant.
2. As the F-ratio based on experimental data is greater than critical F-ratio, the null hypothesis $H_{0,2}$ is rejected. Hence the subject differences are also significant.

***Example* 2:** The following data are C_{max} (mcg/ml) observed following the administration of 3 drug products each to 4 subjects. Test the significance of product differences and subject differences at 5% ls.

Subject	Products		
	A	B	C
1	2	3	5
2	3	3	4
3	3	4	5
4	3	4	4

Solution: This is an example of ANOVA two way classification. The two factors involved are (i) drug products and (ii) subjects. To test the significance of the product differences and subject differences the following two null hypotheses were framed.

Null Hypotheses:

$H_{0,1}$: Product differences are not significant.

$H_{0,2}$: Subject differences are not significant.

Subjects (Blocks)	Products			T_i	T_i^2/K
	A	B	C		
1	2	3	5	10	33.33
2	3	3	4	10	33.33
3	3	4	5	12	48
4	3	4	4	11	40.33
T_i	11	14	18	T = 43	$\Sigma\ (T_i^2/K) = 154.99$
T_i^2/r	30.25	49.00	81	$\Sigma\ (T_i^2/r\ ) = 160.25$	

$$T_{SS} = \Sigma\, X_{ij}^2 - T^2/N$$

T = Total of all values = 24

N = total number of values = 12

$$T^2/N = \frac{(43)^2}{12} = 154.08$$

ΣX_{ij}^2 = sum of squares of all values

Subjects (Blocks)	Products			
	A	**B**	**C**	
1	4	9	25	
2	9	9	16	
3	9	16	25	
4	9	16	16	
X_{ij}^2	31	50	83	$\Sigma X_{ij}^2 = 163$

$$T_{SS} = \Sigma\, X_{ij}^2 - \frac{T^2}{N}$$

$$= 163 - 154.08$$

$$= 8.92$$

$$T_{rSS\ (product)} = \frac{\Sigma T_i^2}{r} - \frac{T^2}{N}$$

$$= 160.25 - 154.08 = 6.17$$

$$Bl_{SS\ (subjects)} = \frac{\Sigma T_i^2}{K} - \frac{T^2}{N}$$

$$= 154.99 - 154.08$$

$$= 0.91$$

$$E_{SS} = T_{SS} - (T_{rSS} + Bl_{SS})$$

$$= 8.92 - (6.17 + 0.91)$$

$$= 1.84$$

ANOVA Table

Source of Variation	df	SS	Mss = (SS/df)	Fratio	$F_{0.05\ (x,x)}$
Total	11	8.91	0.81	-	
Treatment (Products)	2	6.17	3.08	3.08/0.306 = 10.08	$F_{0.05(2,6)}$ = 5.14
Block (subject)	3	0.91	0.30	0.30/0.306 = 0.98	$F_{0.05\ (3,6)}$ = 4.76
error	6	1.84	0.306		

Results:

1. *Testing product differences*: The F ratio based on experimental data is greater than critical F-ratio. Hence the null hypothesis $H_{0,1}$ is rejected. Hence the product differences are significant at 5% ls.

2. *Testing subject differences*: The F ratio based on experimental data is less than critical F-ratio. Hence the null hypothesis $H_{0,2}$ is accepted. Hence the subject differences are not significant at 5% ls.

Example 3: Find the significance of age and gender on the weight of children at 5% ls based on the given data.

Solution: This is an example of ANOVA two way classification. The two factors involved are 1. age 2. gender. Hence the total variance in the data is divided into three components namely variance due to age, variance due to gender and variance due to chance factors.

Null hypotheses:

$H_{0,1}$: Gender effects are not significant

$H_{0,2}$: Age effects are not significant

Gender	Age group						
	1	**2**	**3**	**4**	**5**	T_i	T_i^2/r
Male	8	8	9	10	21	56	627.2
Female	10	11	10	12	17	60	720
T_i	18	19	19	22	38	$T = 116$	$\Sigma\,(T_i^2/r) = 1347.2$
T_i^2/K	162	180.5	180.5	242	722	$\Sigma\,(T_i^2/K) = 1487$	

$$\text{Tss} = \Sigma X_{ij}^2 - T^2/N$$

T = Total of all values

N = total number of values = 10

$$T^2/N = \frac{(116)^2}{10} = 1345.6$$

ΣX_{ij}^2 = sum of squares of all values

Gender	Age group					
	1	**2**	**3**	**4**	**5**	
Male	64	64	81	100	441	
Female	100	121	100	144	289	
X_{ij}^2	164	185	181	244	730	$\Sigma X_{ij}^2 = 1504$

$$\text{Tss} = \Sigma X_{ij}{}^2 \frac{T^2}{N}$$

$$= 1504 - 1345.6$$

$$= 158.4$$

$$\text{Trss}_{\text{(gender)}} = \frac{\Sigma T_i^2}{r} - \frac{T^2}{N}$$

$$= 1347.2 - 1345.6$$

$$= 1.6$$

$$\text{Blss}_{\text{(age)}} = \frac{\Sigma T_i^2}{K} - \frac{T^2}{N}$$

$$= 1487 - 1345.6$$

$$= 141.4$$

$$\text{Ess} = T_{SS} - (T_{rss} + Bl_{ss})$$

$$= 158.4 - (1.6 + 141.4)$$

$$= 15.4$$

ANOVA Table

Source of Variation	df	SS	Mss = (SS/df)	F ratio	$F_{0.05\ (x,x)}$
Total	9	158.4	20.6	-	
Treatment (Gender)	1	1.6	1.6	3.85/1.6 = 2.40	$F_{0.05(1,4)}$ = 5.14
Block (Age)	4	141.4	35.35	35.35/3.85 = 9.18	$F_{0.05\ (4,4)}$ =4.76
Error	4	15.4	3.85		

Result:

(i)   The calculated F ratio is less than critical F ratio at 5% ls and hence the null hypothesis is accepted. Hence the gender is not significant

(ii) The calculated F ratio is greater than critical F ratio at 5% ls and hence the null hypothesis is rejected. Hence the age groups are significant.

***Example* 4:** The following data refers to the dissolution rate of three drug products A, B, C tested by three different methods MI, MII, MIII. Find the significance of product difference and method difference

	Product		
Method	**A**	**B**	**C**
MI	8	7	2
MII	7	5	3
MII	7	5	3

Solution: This is an example of ANOVA two way classification. The two factors involved are (i) Drug product (ii) Method. The total variance in the data is due to three factors namely drug products, testing method and chance factors.

Null hypotheses:

$H_{0,1}$: Product differences are not significant

$H_{0,2}$: Method difference are not significant

Block (Method)	**Products**				
	A	**B**	**C**	T_i	T_i^2/K
MI	8	7	2	17	96.3
MII	7	5	3	15	75.0
MIII	7	5	3	15	75.0
T_i	22	17	8	T= 47	$\sum (T_i^2/K) = 246.3$
T_i^2/r	161.3	96.3	21.3	$\sum (T_i^2/r) = 278.9$	

$$Tss = \Sigma X_{ij}^2 - T^2/ N$$

T = Total of all values

N = total number of values = 9

$$T^2/ N = \frac{(43)^2}{9} = 245.44$$

$\Sigma X_{ij}^2 = $ sum of squares of all values

	Treatment			
	A	**B**	**C**	
MI	64	49	4	
MII	49	25	9	
MIII	49	25	9	
X_{ij}^2	162	99	22	$\Sigma X_{ij}^2 = 283$

$$Tss = \Sigma X_{ij}^2 - \frac{T^2}{N}$$

$$= 283 - 245.4$$

$$= 37.6$$

$$\text{Trss}_{(product)} = \frac{\Sigma T_i^2}{r} - \frac{T^2}{N}$$

$$= 278.9 - 245.4$$

$$= 33.5$$

$$\text{Blss}_{(Method)} = \frac{\Sigma T_i^2}{K} - \frac{T^2}{N}$$

$$= 246.3 - 245.4$$

$$= 0.9$$

$$\text{Ess} = T_{SS} - (T_{rss} + Bl_{ss})$$

$$= 37.6 - (33.5 + 0.9)$$

$$= 3.2$$

ANOVA Table

Source of Variation	df	SS	Mss = (SS/df)	Fratio	$F_{0.05\ (x,x)}$
Total	8	37.6	4.7	-	
Treatment (Product)	2	33.5	16.75	16.75/0.8 = 20.9	$F_{0.05(2,4)} = 6.94$
Block (Method)	2	0.9	0.45	0.8/0.45 = 1.77	$F_{0.05\ (2,4)} = 6.94$
Error	4	3.2	0.8		

Result:

1. *Product difference*: The calculated F ratio based on experimental data is greater than critical F ratio at 5% ls and hence the null hypothesis H_{01} is rejected. Hence the product difference are significant.

2. *Method difference*: The calculated F ratio based on experimental data is less than critical F ratio at 5% ls and hence the null hypothesis H_{02} is accepted. Hence the method difference are not significant.

Example 5: The number of days of hospitalization in a year is classified by age-group and gender to find out if the hospitalization days differ by age and by gender. The data are given in the following table

Days of Hospitalization during a Year

Gender	Age Group				
	20-29	30-39	40-49	50-59	60+
Male	8	8	9	10	21
Female	10	11	10	12	17

The following two null hypotheses can be tested using the data

H_{01}: The means of hospitalization days of all age-groups are equal

H_{02}: The means of hospitalization days of genders are equal.

The results of two –way analysis of variance are summarized in the following Table

ANOVA Table

Source of variation	Sum of squares	Df	Mean squares	F-ratio	Critical F-ratio at 5%
Age-groups	141.400	4	35.350	9.182	$F_{0.05(4,4)}=6.39$
Gender	1.600	1	1.600	0.416	$F_{0.05(1,4)}=7.71$
Residual	15.400	4	3.850		
Total	158.400	9	17.600		

In the case of age groups the calculated F value > the critical F value.

Hence the null hypothesis H_{01} is rejected

The means of the hospitalization days between different age groups are significantly different.

In the case of gender groups the calculated F value < the critical F value

Hence the null hypothesis H_{02} is accepted

The means of the hospitalization days between the gender are not differently different.

ANOVA Three Way Classification

In the ANOVA three way classification the data are divided in three ways based on three factors of variability. For example, when the effects of temperature, pH and catalyst on the yield of a chemical reactions studied, the data obtained are divided in three directions based on the three factors, temperature, pH and catalyst. The total variance present in the data is divided into the following four components.

(i)   Variance due to chance factors

(ii) Variance due to first factor of variability (rows)- temperature

(iii) Variance due to the second factor of variability (columns) - pH

(iv) Variance due to treatment (catalyst or catalyst concentration

The variance quantities estimated due to temperature, pH and catalyst are compared separately each to the chance variance by an F – ratio test. Significant F- ratios indicate the significance of the effects of the three factors involved. Thus three factors can be tested simultaneously for their significance. Data obtained in LSD is analyzed as per ANOVA three way classifications.

Chapter 8

Research Methodologies - III
(Designs for Laboratory Investigations)

Design of experiments' has become an important area in statistics and has been widely used by different researchers in different subjects ranging from life sciences to social sciences including business applications. Generally, once the results of an experiment are available, then the technique of analysis of variance is applied to study about the possible variances relating to the characteristics of interest. In order to gather more information as much as possible from the experiment, it should be well – designed and pre plan. A well design plan is essential in research and it is in this context the scientists usually depend on the help of statisticians. A statistical design is conceived to be a very good plan for the collection and analysis of data. A well - designed experiment is expected to provide valid conclusions with least possible sampling errors. Generally, experiments are conducted to ascertain a fact or to verify the validity of a particular hypothesis.

For example in pharmacy we are interested to know the effects of factors such as temperature, pH and catalyst on the yield of a chemical reaction or to know the effects of different excipients and their concentrations on the

LEARNING OBJECTIVES

- To know
- Importance of experimental design
- Criteria of a good design
- Aim and objectives with model example.
- Principles: Randomization, replication and local control.
- Completely Randomised Design (CRD)
- Randomised Block Design (RBD)
- Latin Square Design (LSD).
- Principles, applications and analysis of data in each case.

dissolution rate of tablets or to compare the bio equivalence of four drug products etc., In all such cases the factors involved or the related hypotheses can be better verified through a well- designed plan of experimentation. The design of experiments helps in the planning of the experiment in gathering as much as possible relevant information and in making statistical analysis.

 ## Criteria of a Good Design

1. The experiment/design must be free from 'bias'. If bias is there in the planning and execution of an experiment, the rules of probability are not applicable. All statistical methods and tests are based on probability rules only. An experiment should give unbiased estimates or results to enable us to apply statistical tests to draw valid conclusions. Bias in the experimental designs is avoided by randomization in the allotment of treatments to the experimental units.

2. *There must be a measure of error*: The estimates or results of an experiment are not true values, they are measures of true values and hence include error always. Error is the difference between true value and estimated value .The significance or other wise of a result is tested by comparing it with error. Hence without estimating error we cannot evaluate the significance of the results. To measure error, an experiment has the repeated a few times. Error is estimated in the form of range, mean error, standard error or fiducial limits.

3. The experiments should have sufficient accuracy to accomplish its purpose. Accuracy is a measure of the difference between true value and estimated value i.e., error. An experiment is said to be giving accurate results when the error is low. Error can be minimised by avoiding technical errors and inter subject variability especially in biological work. Error should be kept minimum to find out the small real significant differences between two products or treatments.

4. *There must be a clearly defined aim and objective*: The aims and objectives are to be clearly defined before planning an experiment or investigation. For this, the investigator should have a thorough knowledge in the area of his investigation. The current status of knowledge in the area of research and scope for further research to generate new knowledge in the area are needed to define objectives and to plan experiments to fulfil the objectives.

There is no real difference between an aim and an objective. Aim points out the general purpose of the study, whereas objectives spell out exactly what one intends to do in the study. The objectives clearly state what the researcher plans to do.

A Model Example: Aim, objectives and plan of an investigation is described below as a model

Project title: QbD Approach for the Formulation Development of Valsartan Tablets

Introduction: QbD is a systematic approach to pharmaceutical product development that begins with predefined objective and involves designing and developing formulations to ensure predefined product quality objectives. QbD approach in formulation development is based on optimization by factorial designs to select the best combinations of excipients and their levels to achieve the desired product quality.

About 95% of all new potential therapeutic drugs (APIs) exhibit low and variable oral bioavailability due to their poor aqueous solubility at physiological pH and consequent low dissolution rate. These drugs are classified as class II drugs under BCS with low solubility and high permeability characters. These BCS class II drugs pose challenging problems in their pharmaceutical product development process. Valsartan, a widely prescribed anti hypertensive drug belongs to class II under BCS classification and exhibit low and variable oral bioavailability due to its poor aqueous solubility. Because of poor aqueous solubility and dissolution rate it poses challenging problems in its tablet formulation development. It needs enhancement in the dissolution rate in its formulation development.

Several techniques such as micronisation, cyclodextrin-complexation, use of surfactants, solubilizers and superdisintegrants, solid dispersion in water soluble and water dispersible carriers, microemulsions and self emulsifying micro and nano disperse systems have been used to enhance the solubility, dissolution rate and bioavailability of poorly soluble BCS class II drugs. Among the various approaches cyclodextrin complexationand use of superdisintegrants such as crospovidone and sodium starch glycolate (Primojel) are simple industrially useful approaches for enhancing the dissolution rate of poorly soluble drugs in their formulation development. Complexation with β-cyclodextrin (βCD) and use of superdisintegrants namely Crospovidone and Primojel are tried in the present study for enhancing the dissolution rate of Valsartan in its formulation development. Formulation of Valsartan tablets with NLT 85% dissolution in 10 min was optimized by 2^2 and 2^3 factorial designs.

Optimization of pharmaceutical formulations involves choosing and combining ingredients that will result in a formulation whose attributes confirm with certain prerequisite requirements. The choice of the nature and qualities of additives (excipients) to be used in a new formulation shall be on a rational basis. The application of formulation optimization techniques is relatively new to the practice of pharmacy. In general the procedure consists of preparing a series of formulations, varying the concentrations of the formulation ingredients in some systematic manner. These formulations are then evaluated according to one or more attributes, such as hardness, dissolution, appearance, stability, taste and so on. Based on the results of these tests, a particular formulation (or series of formulations) may be predicted to be optimal. The optimization procedure is facilitated by applying factorial designs and by the fitting of an empirical polynomial equation to the experimental results. The predicted optimal formulation has to be prepared and evaluated to confirm its quality.

Aim: The aim of the study is optimization of Valsartan tablet formulation with NLT 85% dissolution in 10 min

Objectives

1. Optimization of Valsartan tablet formulation by 2^2 factorial design employing β cyclodextrin (β CD) and two superdisintegrants (crospovidone and Primojel).

2. Optimization of Valsartan tablet formulation by 2^3 factorial design employing β CD, two superdisintegrants (crospovidone and Primojel) and PVP K30.

Plan of the Work

The following is the plan of the work to be carried out:.

1. Procurement of drugs and excipients.

2. *Analytical methods*: Standardization of UV spectrophotometric method for the estimation of valsartan. Validation of the method for linearity, accuracy, precision and interference by the excipients and percent recovery.

3. Studies on optimization of Valsartan tablet formulation by 2^2 factorial design employing β CD and superdisintegrant (two series, one with crospovidone and the other with Primojel).

4. Studies on optimization of Valsartan tablet formulation by 2^3 factorial design employing β CD, superdisintegrant (two series, one with crospovidone and the other with Primojel) and PVP K30.

5. Preparation of Valsartan tablets by Direct compression method as per 2^2 and 2^3 factorial designs employing β cyclodextrin, Superdisintegrants, PVP K30 and evaluation of the tablets prepared for Drug content, Hardness, Friability, Disintegration time and Dissolution rate.

6. Estimation of various dissolution parameters from the dissolution data.

7. Analysis of Dissolution Rates (K_1) and DE_{30}values as per ANOVA of factorial designs.

8. Development of polynomial equation describing the relationship between the response Y (NLT 85% in 10 min) and the variables X_1(Level of superdisintegrant), X_2 (Level of βCD) and X_3 (Level of PVP K 30) based on the experimentally determined data.

9. Developing Optimized formulations of Valsartan tablets based on the polynomial equation and evaluation of the optimized tablet formulations for various parameters including dissolution rate and stability studies.

10. *Evaluation of Drug*: Excipient compatibility by FTIR and DSC studies.

11. Publication of research papers.

Principles of Experimental Designs: There are three important principles underlying the design of experiments. They are:

1. Replication
2. Randomisation
3. Local control

Replication

Replication implies the repetition of the basic treatment or treatment combination under study. If the treatment is applied only once, then it is not possible to understand the treatment effect. On the other hand, if the treatment is applied repeatedly for more number of times, it is possible to understand the variations in treatment effect. If an experiment is replicated for 'r' times, then the standard error of the treatment mean based on 'r' replications is given by $\sigma/\sqrt{r}$ where 'σ' is standard deviation of observations obtained from experimental units. Thus, the standard error of the treatment mean is inversely proportional to square of the number of replications. Accordingly, by increasing the number of replications we can reduce the error variance. Further, the number of replications to be considered in an experiment depends on the knowledge of variability of the experimental material. The number of replications should not be less than 4 in any case.

Advantages

1. Replication averages out the effect of chance factors on experimental units, hence it is possible to improve reliability of estimates through replication.

2. Replication helps in improving precision of a design.

3. It helps in the conduct of the test of significance of the difference between any two treatments.

4.   It helps in the construction of confidence intervals or fiducial limits.

Randomisation

Randomisation is a process that refers to the allocation of treatments to various experimental units such that each treatment has an equal chance of being allocated to the experimental unit. The replication of treatments is made by experimenter so as to reduce experimental error and to get the designs with more precision and accuracy. A critical problem before the experimenter is the allocation of treatments to experimental units, so that on the average each treatment is subjected to equal environmental effect. It is in this context the technique of randomisation helps the experimenter. Further, in the absence of prior information about the variability of experimental material also, the process of randomisation helps the experimenter. Thus, the randomisation ensures equal environmental effect for the all treatments without any bias. However, the randomisation without replication is not useful. Hence, for achieving best results, randomisation principle with adequate number of replications is to be considered.

Principle of Local Control

Local control is division of experimental units into blocks or groups which are more homogeneous and testing all the treatments in each and every block. The purpose of local control is to reduce error due to the variations in the experimental units like fertility of agricultural fields or inter subject variability in biological (animal or human) studies. Inter subject variability in biological/ pharmacological experiments leads large error, which mask the real differences between the products. No two subjects are similar in their physiological and biochemical setup and this gives rise to inter subject variability and error. To avoid the large error, the subjects (human subjects) are divided groups based on the factor of variability like age, body weight, sex etc.., Ultimate division is to take each subject as a block and test all the products in each and every subject. Designs such as Latin Square (LSD) and Randomized Block (RBD) utilize the principle of Local control to reduce error.

Completely Randomised Design (CRD)

The completely randomised design popularly known as CRD, is the simplest of all experimental designs and can be considered as a basic design. All other randomised designs are developed by considering restrictions upon the allocation of the treatments within the experimental area. The CRD is based on the principles of randomisation and replication. In this design, treatments are allocated at random to the experimental units over the entire experimental material. However, the number of repetitions for any one treatment may vary. The CRD may be selected when the variation over all of the experimental material is homogenous. Consequently, CRD is preferred in laboratory experiments, physical, chemical sciences, green house studies etc., as there is much scope for homogeneity of experimental material.

Advantages of CRD

1. The layout of the design is simple.
2. Any number of treatments can be used.
3. Any number of replications can be considered.
4. If some observations are missing on some of the treatment, yet the analysis dose not pose problems.
5. It is possible to reduce the replications corresponding to the treatments for which, the observations are missing and can carry out the analysis of data.
6. The design allows for the maximum number of degrees of freedom in the error sum of squares.

Disadvantages

1. The design is based on the important assumption about the homogeneity of experimental units, which is rarely met in the field experiments. Hence, design is not suitable for field experiments.
2. In this design, if large number of treatments is to be considered then a relatively large amount of experimental material is supposed to be used. Consequently, the assumption of homogeneity may be violated so that experimental error increases.

Applications of CRD:

CRD is applicable in the following types of studies:

1. To study the effect of four temperatures on the yield of a chemical reaction.
2. To study the effect of three binders on the dissolution rate of paracetamol tablets.
3. To study the effect of concentration of lubricant on the dissolution rate of aspirin tablets prepared by direct compression method.
4. To study the effect the reflection properties of four kinds of paints.
5. To study the effect of temperature on the solubility of insoluble drugs.

Analysis of Data

Analysis of the results is carried out by ANOVA one way classification.

Randomised Block Design (RBD)

In this design the principle of local control is used to reduce experimental error. The experimental units or experiments are divided into blocks or groups based on a factor of variability and all the treatments (like products, temperatures, pHs etc.,) are tested in each block or group. In RBD the division of experiments is based on one factor of variability.

Applications

Example 1: To study the effect of temperature and pH on the yield of a chemical reaction.

In this case, the experiments are divided into blocks based on pH. All experiments carried out at particular pH are considered as a 'Block'. If three pHs like 2.0, 4.0, 6.5 are included in the study experiments are divided into three blocks. To know the effect of temperature, the factor temperature has to the studied at two or more levels. If 4 levels of temperature like 40°C, 50°C, 60°C and 70°C are selected, all the four temperatures are to be studied / tested at all the three pHs i.e in all the three blocks. Hence the total number of experiments to be conducted are $3 \times 4 = 12$ experiments.

In this example we are testing the significance of effects of two factors like temperature and pH on the yield of the reaction. Total variation in the results of 12 experiments is due to (i) temperature (ii) pH and (iii) chance factors. The analysis of data is carried out by ANOVA two way classification. The total variation in the data is divided into the following three components

(i)   Variance due to factor temperature

(ii) Variance due to factor pH and

(iii) Variance due to chance factors (chance variation).

The significance of the first two variance components is tested by comparing each to the chance variation by an F-ratio test. Significant F-ratios indicate the significance of the effect of the factors tested. Thus we can evaluate the significance of two factors simultaneously in RBD.

Example 2: To evaluate the effect of 4 binders and 3 disintegrants on the dissolution rate of a tablet formulation.

Example 3: To study the effect of (i) temperature and (ii) aeration rate on the production yield of an Antibiotic fermentation.

Example 4: Bioequivalence testing of four drug products. In this example each subject is considered as a block and all the four drug product are to be tested in each of the four subjects after allowing a washout period of 1 month between testings.

RBD is widely used in pharmacy research involving physical, chemical, biological/ clinical studies.

Latin Square Design (LSD)

In Latin Square Design (LSD) the principles of randomization, replication and local control are used. In this design the experiments or experimental units are divided into blocks or groups based on 2 factors of variability and a third factor can be studied in each block/ group due to the first two factors.The experimental error is much reduced in LSD and more sensitive comparisons can be made. The significance of three factors can be evaluated simultaneously.

The three factors and their levels are arranged in a latin square. The latin square is a square having equal number of rows and columns. One factor and its levels are taken as rows and the second factor and its levels are taken as columns. The levels of third factor are arranged in the latin square in such a way that each level appears only once in each row and each column.

The results of LSD are analysed by ANOVA three way classification. The total variance present in the data is divided into the following four components

   (i)   variance due to factor 1 (rows)

   (ii)   variance due to factor 2 (columns)

   (iii)   variance due to factor 3 (treatment)

   (iv)   chance variation

The significance of the first three variance components is tested by comparing each to the chance variation by an F-ratio test. Significant F-ratios indicate the significance of the effect of the three factors tested. Thus we can evaluate the significance of three factors simultaneously in LSD.

Applications

Example 1: To evaluate the effect of three factors namely (i) temperature (ii) pH and (iii) catalyst concentration on the yield of a chemical reaction. In this example the experimental units or experiments are divided into groups based on two factors, (i) temperature (row wise) and (ii) pH (column wise). The third factor three catalyst concentrations, A (0.1%), B (0.5%), C (1.0%) can be tested each at all temperatures and pHs as shown in the following LSD.

Temp (^{0}C)	pH		
	2	4	6
40	A	B	C
60	C	A	B
80	B	C	A

In the above arrangement all catalyst concentrations (A, B, C) are tested only once in each row and in each column of LSD.

***Example* 2:** Application of LSD in Bioequivalence study of four drug products:

USFDA has prescribed LSD as experimental design for bioequivalence testing of drug products. The performance of the products tested will be influenced by the subject used as well the time period of the study particularly if the study extends over a long period of 6 months to 1 year. In such a case, for bioequivalence study the experiments are divided into groups/ blocks based on the two factors of variability namely (i) subject and (ii) time period giving rise to an LSD. Latin square is a square having equal number of rows and columns. As four drug products are involved in the bioequivalence study, a 4×4 LSD is suitable design. The following is a suitable design for bioequivalence testing of four drug products

	T1	**T2**	**T3**	**T4**
Subject 1	A	**B**	C	D
Subject 2	D	A	B	C
Subject 3	C	D	A	B
Subject 4	B	C	D	A

The interpretation of the above design is as follows:

The experiment marked as **B** involves testing product B in subject 1 at time period 2. All the 4 products are tested only once in each row and in each column i.e once in each subject and once at each time period. The purpose this allotment is to nullify the influence of subjects and time period and hence the experimental error is much reduced.

The results obtained are analyzed as per ANOVA three way classification and we can find out the significance of bioavailability differences of products nullifying inter subject variability and effect of time period.

***Advantages*:**

1. Error is much reduced and hence comparisons are more sensitive.
2. We can also find out the significance of inter subject variability and time effects in the bioequivalence testing of drug products.
3. LSD can be easily replicated to increase the size or replication size (n).
4. More scientific than other designs.
5. LSD can be used in physical, chemical, biological and clinical studies to evaluate the effects of three factors simultaneously.

Chapter 9

Research Methodologies - IV
(Factorial Experiments)

Generally, experiments can be classified into two types, (i) Varietal trails (simple designs) and (ii) Factorial experiments. The CRD, RBD and LSD, which are described earlier corresponds to the varietal or simple trials. In all these three types of experiments, our interest is to study the effects of only one variable i.e.., treatment. Thus, the important objective of those experimental designs is to compare and estimate the effects of a single set of treatments such as different pHs, temperatures, variety of fertilizers, seeds of different quality, different methods of manufacturing, treatment with different drugs, different feeds, etc.., In other words, those experimental designs namely CRD, RBD and LSD restrict the scope of study to only one experimental variable or factor. However, in practise we may be interested to study the simultaneous effect of two or more experimental variables or factors. It is in this context factorial experiments are more useful to the experimenter for the evaluation of combined effect of two or more experimental variables when used simultaneously. Further, it is also possible to examine the interaction effect of variables (factors) i.e.., the combined effect of variables (factors)

LEARNING OBJECTIVES

- To know
- Factorial designs: definitions and applications.
- To study 2^2 and 2^3 factorial designs with worked out examples.

under study. There exists a wide variety of experimental designs in the literature of factorial experiments.

Advantages of Factorial Experiments

1. Factorial experiments are useful to study not only the individual effects of each factor but also their interaction effects.

2. The main effects in factorial experiment are evaluated over a wide range of conditions and information is obtained with the minimum expenditure of resources.

3. All the experimental units are utilised in evaluating effects, resulting in the most efficient use of resources.

4. In factorial experiments, it is possible to save time and experimental material, as time required for the combined experiment is less than the time required for separate experiments.

5. It is possible to secure more information through factorial experiments compare to single factorial experiments.

Disadvantages

1. If large number of treatment combinations is used then the execution of experiment and the statistical analysis becomes more complex.

2. In factorial experiments, increased number of factors will result in increased number of treatment combinations, which in turn increases the block size. Further, increased block size results in increased heterogeneity of the experimental material, which ultimately results in increased error. The increased error effects the precision of experiment.

2^2 Factorial Experiments

Let us consider two factors A and B each at two levels (0, 1) so that there will be $2 \times 2 = 4$ treatment combinations. Let us also suppose that "a" and "b" denote one of the two levels of each corresponding factors and this will be called as second level. Further, let us express the first level of factors A and B by the absence of corresponding letters in treatment combinations. Now the four treatment combinations can be expressed as follows

$a_0 b_0$ or '1' denotes the factors A and B at first level.

$a_1 b_0$ or 'a' denotes the factors A at second level and B at first level.

$a_0 b_1$ or 'b' denotes the factors A at first level and B at second level.

$a_1 b_1$ or 'ab' denotes the factor A and B both at second level.

Let us suppose that the factorial experiment with 4 treatment combinations is taken up with "r" plots i.e.., replications. Let [1], [a], [b] and [ab] denote the total yield of the "r" plots receiving the treatments 1, a, b and ab respectively. Also let us consider mean values of yield (obtained by dividing the total yield with r units) be given by (1), (a), (b) and (ab). Further, let us suppose that A, B and AB denotes the main effects due to the factors A & B and their interactions AB respectively.

For example when two factors such as temperature and pH are influencing the yield of chemical reaction, it is quite possible that in addition to their individual effects, they may produce a combined or interaction effect on the yield of the chemical reaction. Two factors are said to be interacting when the effect of one factor is different at different levels of second factor.

For example, A and B are two factors like temperature and pH. To evaluate their effects, each factor has to be studied minimum at two levels. Let a_0 and b_0 are the first levels of the two factors A and B and a_1 and b_1 are the second levels of the two factors. Then, the effect of factor A i.e $(a_1 - a_0)$ at b_0 is equal to $(a_1 - a_0)$ at b_1 which means the effect of A is same at two levels of the other factor B, then there is no interaction effect between two factors A and B. Alternatively, when $(a_1 - a_0)$ at b_0 is different from $(a_1 - a_0)$ at b_1 which means the effect of factor A is different at the two levels of the other factor B, then the two factors are said to be interacting.

Selected combinations of the factors and their levels are taken as treatments in the factorial experiments to evaluate the individual and combined effects of the factors involved. To evaluate the effect, each factor has to be studied minimum at two levels. 2^2 factorial experiments means two factors each at two levels. The treatments are obtained by expanding 2^2 which is equal to four. Hence a 2^2 experiment require four treatments based on the selected combinations of the factors and their levels of the two factors.

For a 2^2 experiment, the following are the four treatments required.

1 (Both the factors A and B at first level)

a (Factor A at second level and Factor B at first level)

b (Factor A at first level and Factor B at second level)

ab (Both the factors A and B at second level)

The treatments are then tested as per a suitable experimental design such as CRD, RBD and LSD. Each treatment has to be replicated a suitable number of times to enable us to estimate the error involved for carrying out test of significance.

The data obtained in a factorial study are analysed as per ANOVA one way classification to find out the significance of the treatments (a,b,ab) involved. The individual and combined effects of the two factors are estimated by Yates' method and are subjected to F-ratio test in the analysis of variance (ANOVA).

Analysis of Data

Considering CRD or RBD or LSD can carry out the factorial experiments as usual and hence, the analysis part is also similar to that already discussed in those designs earlier with slight modification relating to treatment sum of squares is divided into three orthogonal components with 1 degree of freedom. The sum of the squares due to factorial effect A, B and AB can be obtained by multiplying the square of the factorial effect by a suitable quantity. Factorial effect totals are given by

$$[A] = [ab] - [b] + [a] - [1]$$

$$[B] = [ab] + [b] - [a] - [1]$$

$$[AB] = [ab] - [a] - [b] + [1]$$

Total sum of squares due to any factorial effect is obtained on multiplying the square of the effect total by the factor (1/4r) where r is the replication. Thus, the sum of squares due to factorial effect each with 1 degrees of freedom can be written as

Sum of squares due to main effect of $A = [A]^2/4r$

Sum of squares due to main effect of $B = [B]^2/4r$

Sum of squares due to main effect of $AB = [AB]^2/4r$

The ANOVA summary for 2^2 experiments with' r' replication considering fixed effect model of RBD can be written as follows

Source of Variation	Degree of freedom	Sum of Squares	Mean Sum of Squares	F – ratio
Blocks	r-1	RSS	MRS=RSS/(r-1)	F_b=MRS/MES
Treatments	2^2-1	TSS	MTS=TSS/(2^2-1)	F_T=MTS/MES
Main effect A	1	SSA	MSA=SSA/1	F_A=MSA/MES
Main effect B	1	SSB	MSB=SSB/1	F_B=MSB/MES
Main effect AB	1	SSAB	MSAB=SSAB/1	F_{AB}=MSAB/MES
Error	(r-1) (2^2-1)	SSE	MES=SSE/(r-1)(2^2-1)	-------

The calculated value of F-ratio, corresponding to blocks and treatments is greater than the critical value i.e. F_α[(r-1), 3(r-1) and F_α [(t-1), 3(r-1) then reject the null hypothesis at a level of significance, implying that block effect as well as treatment effect are not equal. Otherwise accept null hypothesis. Similarly, if calculated value of F corresponding to main effect and interaction effects i.e. A, B and AB is greater than critical value i.e.

$F_{\alpha\ [1,3(r-1)]}$ then reject null hypothesis implying that the factorial effects (individual and interaction effects) are significant at the chosen level of significance.

Computation of Factorial Effect Totals-Yates' Method

Yates method provides frame work for the computation of factorial effect total for 2^n factorial experiment. For 2^2 factorial experiment the following procedure is adopted to compute the factorial total effects.

Step 1: Enter the treatment combination in the first column such that starting with the treatment combination 1 followed by each factor a,b and then treatment combinations i.e, ab.

Step 2: Enter total yield corresponding to each and every treatment combination in column 2 for all replicates i.e., [1],[a],[b],[ab].

Step 3: In third column, against the treatment combination '1' enter the sum of total yield corresponding to 1 and a i.e., [1]+[a], against the treatment combination a enter the sum of total yield corresponding to b and ab i.e [b]+[ab], against the treatment combination b enter the difference of total yield corresponding to a and 1 i.e., [a]-[1], against the treatment combination 'ab' enter the difference of total yield corresponding to ab and b i.e.,[ab]-[b].

Step 4: In column 4 follow the procedure outline in step 3 so that write [1]+[a]+[b]+[ab] ,[ab]-[b]+[a]-[1], [ab]+[b]-[a]-[1],[ab]-[b]-[a]+[1] against 1, a,b, and ab which gives corresponding factorial effect total.

Above steps are shown in the following Table:

Treatment combination(1)	Total yield from all replication(2)	(3)	Factorial effect Total (4)
[1]	[1]	[1]+[a]	[1]+[a]+[b]+[ab]
[a]	[a]	[b]+[ab]	[ab]-[b]+[a]-[1]
[b]	[b]	[a]-[1]	[ab]+[b]-[a]-[1]
[ab]	[ab]	[ab]-[b]	[ab]-[b]-[a]+[1]

Example 1:

To evaluate the effects of β CD (β cyclodextrin) and a surfactant (SLS) on aqueous solubility of ibuprofen, a poorly soluble drug

This study can be planned as 2^2 factorial experiment. Factor A (β CD) and factor B (SLS) may have their individual effects as well as combined effect in enhancing the solubility of the drug. In this experiment the two levels of β CD are taken as 0 and 5 mM and two levels of SLS are 0 and 2 %. Hence the four treatments involved in the study are as follows

1 (both the factors at first level i.e. water alone)

a (water containing factor A at second level i.e. water containing 5mM β CD)

b (water containing SLS at second level i.e. water containing 2% SLS)

ab (both the factors at the second level i.e. water containing 5mM of β CD and 2% SLS)

The solubility of ibuprofen has to be determined in the above four fluids as per 2^2 factorial study. Each solubility determination is to be replicated a suitable number of times like (n = 4). The data obtained (4 × 4 = 16 solubility measurements) are then analysed as per ANOVA one way classification and also as per Yates method to know the significance of the individual and combined effects of the two factors involved i.e. β CD and SLS in enhancing the solubility of ibuprofen.

***Example* 2:** Two study the Effects of βCD, Poloxamer 407 and PVP on the Solubility of

Pioglitazone Design: 2^3 Factorial Study.

This is an example of 23 factorial design in which three factors are evaluated each at two level

The three factors involved in the study are

Factor A is β CD (the two levels are 0, 5 mM)

Factor B is Poloxamer 407 (the two levels are 0, 2%)

Factor C is PVP(the two levels are 0, 2%)

For a 2^3 design, eight selected combinations of the factors and their levels are needed. They are as follows. 1, a, b, ab, c, ac, bc and abc.

The solubility of Pioglitazone in the selected 8 fluids is determined each four time s(n = 4). The data and analysis is shown in the following tables.

Effects of βCD, Poloxamer 407 and PVP on the Solubility of Pioglitazone Design: 2^3 Factorial Study

Solubility (mg/ml), A: β CD (0, 5 mM) B: Poloxamer 407 (0, 2%)C: PVP (0, 2%)

1	a	b	ab	c	ac	bc	abc
0.128	0.432	0.493	1.624	1.224	0.776	1.749	2.202
0.112	0.408	0.574	1.681	1.192	0.873	1.572	1.495
0.120	0.416	0.554	1.745	1.103	0.921	1.725	1.653
0.128	0.416	0.489	1.713	1.139	0.881	1.620	1.766
Mean 0.122	0.418	0.527	1.691	1.165	0.863	1.667	1.779
SD 0.008	0.010	0.043	0.052	0.054	0.062	0.085	0.303
Folds increase	3.422	4.32	13.85	9.54	7.06	13.65	14.57

ANOVA Table

Source of variation	df	ss	mss	F ratio	significance
Total	31	11.93	0.385		
treatment	7	11.60	1.657	120.152	$P < 0.01$
error	24	0.33	0.0138		

Analysis as per Yates method to evaluate the main and interaction effects

Yates method	Total			Ti	Ti^2	$Ti^2/32$
1	0.49	2.16	11.03	32.92	1083.99	33.87468
a	1.67	8.87	21.89	5.08	25.8064	0.80645
b	2.11	8.11	5.84	12.39	153.413	4.794156
ab	6.76	13.78	-0.76	5.13	26.27588	0.821121
c	4.66	1.18	6.71	10.86	117.8962	3.684255
ac	3.45	4.65	5.673	-6.59	43.48084	1.358776
bc	6.67	-1.21	3.47	-1.04	1.0816	0.0338
abc	7.12	0.45	1.66	-1.81	3.283344	0.102605

Source of variation	df	ss	mss	F ratio	significance
a	1	0.81	0.80645	58.46621	$P < 0.01$
b	1	4.79	4.794156	347.5679	$P < 0.01$
ab	1	0.82	0.821121	59.52984	$P < 0.01$
c	1	3.68	3.684255	267.102	$P < 0.01$
ac	1	1.36	1.358776	98.50888	$P < 0.01$
bc	1	0.03	0.0338	2.450441	$P > 0.05$
abc	1	0.10	0.102605	7.438646	$P < 0.01$
error	24	0.33	0.013795		
$F_{0.01}(7,24)=3.50,$		$F_{0.01}(1,24)=7.82,$			
$F_{0.05}(7,24)=2.4$		$F_{0.05}(1,24)=4.26.$			

Result: All the main and combined (interaction) effects of the three factors are highly significant ($p < 0.01$) except the combined effect of bc (polaxomer-PVP)

Chapter 10

Research Methodologies - V
(DoE and Optimization)

In Pharmacy word "optimization" is found in the literature referring to any study of formula. In development projects pharmacist generally experiments by a series of logical steps, carefully controlling the variables and changing one at a time until satisfactory results are obtained. This is how the optimization done in pharmaceutical industry.

Optimization is defined as follows: "Choosing the best element from some set of available alternatives". It is the process of finding the best way of using the existing resources while taking in to the account of all the factors that influences decisions in any experiment. The objective of designing quality formulation is achieved by various Optimization techniques like DoE (Design of Experiment).

The term FbD (Formulation by Design) & QbD (Quality by Design) indicates that quality in the product can be built by using various techniques of DoE (Design of Experiment).

This FbD has replaced the OVAT (one variable at a time) strategy for Optimization completely.

LEARNING OBJECTIVES

- QbD, DoE and Optimization techniques, their significance and applications
- Types of optimization techniques
- Plackett burman design with a worked out example
- Factorial designs
- Optimization by factorial designs and multiple regression
- Applications and case studies on factorial designs

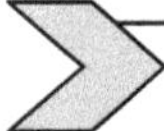

Quality by Design (QbD)

The pharmaceutical Quality by Design (QbD) is a systematic approach to development that begins with predefined objectives and emphasizes product and process understanding and process control, based on sound science and quality risk management. Quality by Design (QbD) is emerging to enhance the assurance of safe, effective drug supply to the consumer, and also offers promise to significantly improve manufacturing quality performance.

Application of QbD in Pharmaceutical Industry

Quality refers to product free of contamination and delivers the therapeutic benefit promised in the label to the consumer. The Quality of the pharmaceutical product can be evaluated by in vivo or in vitro performance tests "QbD" assures in vitro product performance and In vitro product performance provides assurance of in vivo product performance. "Hence QbD relate to Product Performance".

Benefits for industry include, better understanding of the process, less batch failure, more efficient and effective control of change, return on investment / cost savings, provides opportunities for more flexible regulatory approaches, manufacturing changes within the approved design space without further regulatory review, reduction of post-approval submissions, better innovation due to the ability to improve processes without resubmission to the fda when remaining in the design space.

Design of Experiment (DoE)

DOE is a mathematical tool for systematically planning and conducting scientific studies that change experimental variables together in order to determine their effect on a given response. It makes controlled changes to input variables in order to gain maximum amounts of information on cause and effect relationships with a minimum sample size for optimizing the formulation.

There are mainly four steps associated with DoE:

1. The design of the experiment (By using various models)
2. The collection of the data
3. The statistical analysis of the data and
4. The conclusions reached and recommendations made as a result of the experiment.

In Optimization various types of models used from preliminary screening of factors to select their levels and for finally to study their effect. It depends upon the formulator to choose a suitable model for study and help in minimizing the experimenting time.

 # Types of Experimental Design Methods

There are various types of Experimental design methods available out of which the method we have to use depends upon the resources we have and what we want to study.

Screening Designs are used for identify the important factor and their level which affect the quality of formulation. Screening designs generally support only the linear responses.

Response Surface Designs are used when we required exact image of response, estimating interaction and even quadratic effects. Response surface designs generally support non linear and quadratic response and capable of detecting curvatures

Factorial Designs

Factorial designs (FDs) are very frequently used response surface designs. A factorial experiment is one in which all levels of a given factor are combined with all levels of every other factor in the experiment. These are generally based upon first-degree mathematical models. Full FDs involve studying the effect of all the factors *(k)* at various levels (x), including the interactions among them, with the total number of experiments being x^k. If the number of levels is the same for each factor in the optimization study, the FDs are said to be *symmetric,* whereas in cases of a different number of levels for different factors, FDs are termed *asymmetric."*

When we study three factors at two level 2^3 the total number of run will be = 08 &

When we study two factors at three level 3^2 the total number of run will be = 09

Fractional Factorial Design (FFD)

Fractional factorial design is generally used for screening of factor. This design has low resolution due to less number of run. Although these designs are economical in terms of number of experiments, the ability to distinguish some of the factor effects is partly sacrificed by reduction in the number of experiments.

Plackett-Burman Designs (Hadamard Designs)

Plackett-Burman designs (PBD) are special two-level FFDs used generally for screening of factors. This design is generally used when we want to screen high number of factors (11-47) if we want to study the effect of 7 factors then we have to show four dummy factors. The interpretations of results in FFD, Plackett-Burman designs & Taguchi design are drawn with the help of Pareto chart and Half normal plot.

Central Composite Design (Box-Wilson Design)

For nonlinear responses requiring second-order models, central composite designs (CCDs) are the most frequently employed. A two-factor CCD is identical to a 3^2 FD with rectangular experimental domain at $\alpha = \pm1$, on the other hand, the experimental domain is spherical in shape for $\alpha = \sqrt{2} = 1.414$.The CCD is quite popular in response surface optimization during pharmaceutical product development.

Box-Behnken Designs

A specially made design, the Box-Behnken design (BBD), requires only three levels for each factor -l, 0 and +1. It employing 15 experiments run with three factors at three levels. It is economical then CCD because it requires less number of trials.

Taguchi Design

Taguchi refers to experimental design as "off-line quality control" because it is a method of ensuring good performance in the development of products or processes." It is also used for screening of factors and it provides 8 experimental runs for 7 factors.

Mixture Design

Mixture designs are used when the characteristics of the finished product (Drug delivery system) usually depend not so much on the quantity of each substance present but on their proportions. The sum total of the proportions of all the excipients is unity, and none of the fractions can be negative. Therefore, the levels of different components can be varied with the restriction that the sum total should not exceed one.

Model Development for Optimization

A model is an expression defining the quantitative dependence of a response variable on the independent variables. Usually, it is a set of polynomials of a given order or Degree. From this polynomial equation we calculate the coefficient with the help of Principal of MLRA (Multiple

Linear Regression Analysis). By the help of software we can also study here the effect of excipients, their interaction study, 3D Response plot, Contour Plot etc. In screening design with the help of half normal plot and Pareto chart we can find out easily the main factor and their level.

From the models thus selected, optimization of one response or the simultaneous optimization of multiple responses needs to be optimized graphically, numerically and by using Brute force search technology.

(a) *Graphical Optimization*: Graphical optimization deals with selecting the best possible formulation out of a feasible factor space region. To do this, the desirable limits of response variables are set, and the factor levels are screened accordingly by the help of overlay plot.

(b) *Brute-force search (Feasibility and Grid search)*: Brute-force search technique is the simple and exhaustive search optimization technique. It checks each and every single point in the function space. Herein, the formulations that can be prepared by almost every possible combination of independent factors and screened for their response variables. Subsequently, the acceptable limits are set for these responses, and an exhaustive search is again conducted by further narrowing down the feasible region. The optimized formulation is searched from the final feasible space (termed as grid search), which fulfills the maximum criteria set during experimentation.

(c) *Numerical Optimization*: It deals with selecting the best possible formulation out of a suitable factor. To do this, the desirable limits of response variables are set, and the factor levels are displayed by the software.

 ## Validation of Model Developed

The predicted optimal formulation (Check point) is prepared as per optimum factor level and the responses evaluated. On comparison of results of observed and predicted response conclusion will be drawn for model validation.

Software for Designs and Optimization

Many commercial software packages are available which are either dedicated to experimental design alone or are of a more general statistical type.

Software dedicated to experimental designs include DESIGN EXPERT, ECHIP, MULTI-SIMPLEX, NEMRODW, Software for general statistical nature, SAS, MINITAB, SYSTAT, GRAPHPAD PRISM.

Design Expert

Optimizing pharmaceutical formulations and processes; allows screening and study of influential variables for FD, FFD, BBD, CCD, PBD and mixture designs; provides 3D plots that can be rotated to visualize the response surfaces and 2D contour maps; numerical and graphical optimization.

Mini Tab

Powerful DoE software for automated data analysis, graphic and help features, MS-Excel compatibility, includes almost all designs of RSM

JMP

DoE software for automated data analysis of various designs of RSM, graphic and help features

CARD

Powerful DoE software for automated data analysis, includes graphic and help features

DoE PRO XL & DoE KISS

MS-Excel compatible DoE software for automated data analysis using Taguchi, FD, FFD and PBD. The relatively inexpensive software, DoE KISS is, however, applicable only to single response variable.

MATREX

Excel compatible optimization software with facilities for various experimental designs and Taguchi design.

Cornerstone™

DoE software with features for executing various experimental designs

ECHIP

Used for designing and analyzing optimization experiments

GRG2

Mathematical optimization program to search for the maximum or minimum of a function with or without constraints

DoE PC IV

Used for designing the optimization experiments

STATISTICA

ANN-based software based on GRN technique

NEMROD@

Suitable for FDs and CCDs, has features for numerical optimization and graphic outputs.

MODDE

Suitable for response surface modeling and evaluation of fitting of model.

SPSS

Comprehensive statistical software with facilities for implementing experimental designs.

OMEGA

Only for mixture designs; only program that supports multicriterion decision making by Pareto- optimality, up to six objectives and has various statistical functions

DoE WISDOM

Supports designs for screening, D-optimal, Taguchi and user defined designs, also options are available for pare to optimality charts.

COMPACT

Optimization software for systematic DoE and response surface methodology studies with state-of-art mathematical search techniques.

OPTIMA

Generates the experimental design, fits a mathematical equations to the data and graphically depicts response surfaces

XSTAT

Aids in selection of an experimental design, has modules for numerical optimization and graphic outcomes

FACTOP

Aids in the optimization of formulation using various FDs, and other designs through development of polynomials and grid search; includes computer-aided-education module for optimization

Choice of Computer Software Package

When selecting DoE software, it is important to look for not only a statistical engine that is fast and accurate but also the following:

1. A simple graphic user interface (GUI) that's intuitive and easy-to-use.

2. A well-written manual with tutorials to get you off to a quick start.

3. A wide selection of designs for screening and optimizing processes or product formulations.

4. A spreadsheet flexible enough for data entry as well as dealing with missing data and changed factor levels.

5. Graphic tools displaying the rotatable 3-D response surfaces, 2-D contour plots, interaction plots and the plots revealing model diagnostics.

6. Software that randomizes the order of experimental runs.

7. Randomization is crucial because it ensures that "noisy" factors will spread randomly across all control factors.

8. Design evaluation tools that will reveal aliases and other potential pitfalls.

9. After-sales technical support, online help and training offered by manufacturing vendors.

Screening Designs- Plackett Burman Method

Screening design is a method of identification of main factors (variables) among many variables, with the use of less experimental runs.

Examples are screening designs are plackett burman method and fractional factorial designs. These are used as first step (preliminary investigation) in the development of formulation. The most popular screening design is plackett burman design. All screening designs are fractional factorial designs, but there is a slight difference. It assumes that the interaction between variables is negligible and main variable effects are highlighted. Most interaction terms are confounded with the prime variables of interest. Screening methods do not estimate interactions but fractional factorial designs can be used for estimating the interactions. Since analysis of interactions are not possible , further analysis is attempted using more complete designs.

Confounding implies that the effects of two or more treatments are inseparable and therefore the source of any one effect cannot be determined.

Plackett Burman Designs

Plackett burman designs are available in four or multiples of variables.

Number of runs = number of variables + one trial extra (base trial).

For three variables, four experimental runs need to be conducted. Three variables of screening is not preferred because the variance will be very less and result cannot be significant. For example 11 variable plackett burman design, 12 trials (runs) are employed. Similarly in a 15 variables screening design, 16 trials are employed. If the no. of variables is short of the required number, then dummy variable needs to be included. For the evaluation of statistics, it is necessary to conduct replicate analysis.

The designs of a few sets (7 variables, 11 variables and 15 variables) are given. The following rules are considered for writing the design

- Columns represents factors.
- Rows contain process runs
- Two levels, +ve and –ve , are only considered for screening.
- The signs go across down (from right to left hand side) to the next run.
- The first level will be the last level in the subsequent row.
- The last row has all negatives

Table 10.1 Plackett Burman Design with Different Variables.

S.No	X_1	X_2	X_3	X_4	X_5	X_6	X_7	X_8	X_9	X_{10}	X_{11}	X_{12}	X_{13}	X_{14}	X_{15}
8	+	+	+	-	+	-	-								
12	+	+	-	+	+	+	-	-	-	+	-				
16	+	+	+	+	-	+	-	+	+	-	-	+	-	-	-

Applications

When the number of variables is high, screening design is a method of choice. Examples are: fermentation of antibiotics, lactic acid, polymers and other factors in the coating of tablets, formulation and process parameters optimization in the production of tablets.

Advantages

If response (example is coating process) is not well defined, Plackett burman method can be used, example is effluent treatment. If the effects or factors are unknown, screening design can be applied. The number of experiments is fewer compared to parametric methods.

Disadvantages

Screening method does not identify the interactions.

Example: Plackett Burman design is employed for screening the following inorganic nitrogen sources for finding the best for maximization of the production yield of an antibiotic. The eleven ingredients screened include 1) ammonium hydroxide 2) ammonium oxalate 3) ammonium dihydrogen phosphate 4) ammonium chloride 5) ammonium nitrate 6) ammonium hydrogen phosphate 7) ammonium sulphate 8) potassium nitrate 9) ammonium acetate 10) ferric ammonium sulphate and 11) sodium nitrate. The two levels in each case are 0 and 0.3 %.

Solution:

1. In this case the number of experiments to be conducted is 12 for 11 variables (number of variables +1). The same serial number is followed for the nitrogen sources.

2. The first trial (or run) must be compulsorily written as given in table 1. For the remaining trials, the signs are carried down diagonally. The level of X_1 is carried to the X_2 of the next trail. In the 12th trial (run), all values are negative (or lower level).

3. Placket Burman design for this experiment is shown in table 10.2.

Table 10.2 Plackett Burman Design.

Expt No	Ingredients										
	X_1	X_2	X_3	X_4	X_5	X_6	X_7	X_8	X_9	X_{10}	X_{11}
1	+	+	-	+	+	+	-	-	-	+	-
2	+	-	+	+	+	-	-	-	+	-	+
3	-	+	+	+	-	-	-	+	-	+	+
4	+	+	+	-	-	-	+	-	+	+	-
5	+	+	-	-	-	+	-	+	+	-	+
6	+	-	-	-	+	-	+	+	-	+	+
7	-	-	-	+	-	+	+	-	+	+	+
8	-	-	+	-	+	+	-	+	+	+	-
9	-	+	-	+	+	-	+	+	+	-	-
10	+	-	+	+	-	+	+	+	-	-	-
11	-	+	+	-	+	+	+	-	-	-	+
12	-	-	-	-	-	-	-	-	-	-	-

The following are the resposences (yield of antibiotic observed in the 12 runs experiments) conducted

Trials	1	2	3	4	5	6	7	8	9	10	11	12
Yield, units/g	1576	1456	1744	1720	1552	1424	1508	1680	1616	1464	1504	1464

The method of analysis involves the same procedure as done in factorial design. These are $\Sigma(Y_iX_i)$; $\Sigma(X_i^2)$; $B(=\Sigma(Y_iX_i)/\Sigma(X_i^2))$; $SS(=\Sigma(Y_iX_i)^2/(X_i^2))$; contribution of SS in percentage etc. The coefficients and their contributions are in given table 10.3.

***Results*:** Based on the results given in table 2 the following conclusions are be drawn.

1. Ammonium oxalate is the best nitrogen source (%SS=32.64).

2. Ferric ammonium sulphate is also a significant source (%SS=22.62).

3. Ammonium dihydrogen phosphate ($NH_4H_2PO_4$) also has significant effect on the yield (%SS= 11.66).

4. The other 8 sources did not influence on the yield (response). One or more among the three can be selected and included in the media for the effective yield. Further, factorial design is adopted for optimizing the concentration of the ingredients.

Table 10.3 Analysis of Data.

S.No	Nitrogen Source	Coefficient	%SS	Remarks
1	Ammonium hydroxide	-27.0	6.68	
2	Ammonium oxalate	59.67	32.64	Significant
3	Ammonium dihydrogen phosphate	35.67	11.66	Significant
4	Ammonium chloride	1.67	0.03	
5	Ammonium nitrate	-16.33	2.44	
6	Ammonium hydrogen phosphate	-11.67	1.25	
7	Ammonium sulphate	-19.67	3.54	
8	Potassium nitrate	21.0	4.04	
9	Ammonium acetate	29.67	8.07	
10	Ferric ammonium sulphate	49.67	22.62	Significant
11	Sodium nitrate	27.67	7.02	

Optimization by Factorial Designs and Multiple Regression

(A Novel Approach for Pharmaceutical Formulation Development)

The word "Optimize" means to make as perfect, effective or functional as possible. Optimization of product or process is determination of experimental conditions resulting in its optimal performance. Optimization has been defined as the implementation of systemic approaches to achieve the best combination of product and/or process characteristics under a given set of conditions.

With respect to the drug formulations or pharmaceutical process, optimization is a phenomenon of finding "the best" possible composition or operating conditions. Although several optimization procedures are available to the pharmaceutical scientist, in general the procedure consists of preparing a series of formulations, varying the concentrations of formulation ingredients in some systemic manner. These formulations are then evaluated according to one or more attributes, such as hardness, dissolution, appearance, stability, taste and so on. Based on the results of these tests, a particular formulation (or series of formulations) may be predicted to be optimal.

Optimization of pharmaceutical formulations involve choosing and combining ingredients that will result in formulation whose attributes conform to certain pre requisite requirements. The choice of the nature and quantities of additives (or) excipients to be used in a formulation has to be based on some rational. The optimization techniques will help in fixing the quantities or levels of the excipients, Optimization techniques are relatively new to the practice of pharmacy. In general the traditional procedure consists of preparing a series of formulation, varying the concentrations of the formulation ingredients in some systemic manner. These formulations were then evaluated according to one or more attributes such as hardness, dissolution, appearance, stability, taste and so on. Based on the results of these tests a particular formulation or series of formulations may be predicted to be optimal .the predicted optimal formulation has to be prepared and evaluated to conform its quality. The formulation is generally optimized according to a single attribute.

Optimization by Factorial Design

The modern approach for optimization is through the use of statistical techniques. Optimization using factorial designs is an efficient technique used in formulation optimization.

The optimization procedure is facilitated by construction of a mathematical equation that describes the experimental results as a function of the factor levels. A polynomial equation can be constructed by multiple regression technique in the case of a factorial design where the coefficients in the equation are related to effects and interactions of the factors.

The equation constructed from an 2^n factorial experiment is as follows:

$$y = \beta_0 + \beta_1 x_1 + \beta_2 x_2 + \beta_3 x_3 + \ldots + \beta_{12} x_1 x_2 + \ldots + \beta_{123} x_1 x_2 x_3$$

Where y is the measured response, x_i is the level of the ith factor, β_1, β_2, β_3....... represent coefficients computed from the responses of the formulations in the design and β_0 represent intercept.

Full Factorial Design (FFD)

Factorial experiments with two-level factors are used widely because they are easy to design, efficient to run, straightforward to analyze, and full of information. A full factorial design contains all possible combinations of a set of factors. This is the most fool proof design approach, but it is also the most costly in experimental resources. The full factorial designer supports both continuous factors and categorical factors with up to nine levels.

Factorial designs with only two-level factors have a sample size that is a power of two (specifically 2^f where f is the number of factors). When there are three factors have a sample size that is a power of three.

$$N = L^k$$

Where k = number of variables, L = number of variable levels, N = number of experimental trials, for example, in an experiment with three factors, each at two levels, we have eight formulations, a total of eight responses

Case Studies of Optimization Using Factorial Design

Case Study 1: **Formulation of Combined Drug Products**

A 2^2 factorial experiment was designed to develop a combination drug product to obtain the dose of each drug which would result in an optimal response. For this purpose, the 2 levels selected for drug A (x_1) are 5 mg and 10 mg and for drug B (x_2) the two levels are 50 mg and 100 mg. This study is an example of a 2^2 factorial study and involve four formulations with selected combinations of the two levels of drug A and drug B. The four formulations as per 2^2 factorial design are prepared and the response (y) i.e. time to reach anaesthesia in minutes is measured with each formulation.

The formulations as per 2^2 factorial design, their responses (y) observed and potency transformations for developing the polynomial response equation are shown in the following table 10.4.

Table 10.4

Formulation Code	Potency (mg)		Response Time (min) (Y)	Potency Transformed			Response Multiples for Determining Coefficients of Response Equation		
	A (X1)	B (X2)		A (X1)	B (X2)	AB (X1X2)	X1Y	X2Y	X1X2Y
1	5	50	9.7	-1	-1	+1	-9.7	-9.7	+9.7
A	10	50	7.2	+1	-1	-1	+7.2	-7.2	-7.2
B	5	100	8.4	-1	+1	-1	-8.4	+8.4	-8.4
AB	10	100	4.1	+1	+1	+1	+4.1	+4.1	+4.1
						Σ	-6.8	-4.4	-1.8
							$\beta_1=$ -6.8/4 = -1.7	$\beta_2=$ -4.4/4 = -1.1	$\beta_{12}=$ -1.8/4 = -0.45

The polynomial response equation to be developed is of the type

$$y = \beta_0 + \beta_1 x_1 + \beta_2 x_2 + \beta_3 x_3 + \ldots\ldots + \beta_{12} x_1 x_2 + \ldots\ldots + \beta_{123} x_1 x_2 x_3$$

The polynomial equation describing the relationship between the response (y) and the variables x_1 and x_2 based on the observed data was found to be

$$y = 7.35 - 1.7(x_1) - 1.1(x_2) - 0.45(x_1, x_2)$$

Based on the above relationship the optimized formulation with response (y) as 5 min shall contain +0.5 of A(8.75 mg) and +1 of B (100 mg).

Transformed potency = X - Average of two levels / half of the difference of the two levels

For drug A, 0.5 = (X-7.5)/2.5 = 8.75 mg

For drug B, 1.0 = (X-75)/25 = 100 mg

This combination of drug A and B is the optimized formulation which would produce anaesthesia in 5 min. Hence the optimized combined drug formulation should contain 8.75 mg of drug A and 100 mg of drug B.

Case Study 2: **Optimization of Diclofenac SR Tablet Formulation by Factorial Design**

The study is to design diclofenac SR tablets employing a combination of HPMC K 100 M (hydrophilic polymer) and ethyl cellulose (lipophilic polymer) for better controlled release. Diclofenac SR tablet formulation was optimized by 2^2 factorial design. Diclofenac SR tablets were formulated employing the selected combinations of HPMC (Factor A) and EC (Factor B) as per 2^2 factorial study and prepared by wet granulation method. The SR tablets were evaluated for drug release kinetics and mechanism. For optimization, time for 50 % release (T_{50}) was taken as response (Y) and the percent of HPMC as X1 and percent of EC as X2. The polynomial equation describing the relationship between the response Y and the variables X1 and X2 based on the observed data was worked out.

The polynomial equation describing the relationship between the response Y (T_{50}) and the variables X_1 (% HPMC) and X_2 (% EC) based on the observed data was found to be

$$Y = 2.95 + 1.05\,X_1 - 0.25\,X_2 - 1.75\,(X_1 X_2).$$

Based on the above polynomial equation the optimized diclofenac SR tablets with a T_{50} of 4 hours could be formulated employing 50 % HPMC and 5.5 % ethyl cellulose as release retarding polymers. The optimized SR formulation prepared gave slow release of diclofenac over 12 h with a T_{50} of 4 h indicating validity of the optimization technique employed. Diclofenac release from the optimized SR formulation was diffusion controlled and release was by non-fickian (anomalous) diffusion mechanism. Based on pharmacokinetics, diclofenac SR tablets for b.i.d administration should contain a total dose of 100 mg of diclofenac and the desired release rate (K_0) is 8.66 mg/h. The drug release rate of optimized SR tablets formulated was found to be   8.54 mg/h, which is very close to the theoretical desired release rate.   Hence the optimized formulation is considered as the best diclofenac SR formulation developed.

Case Study 3: **Optimization of Valsartan Tablet Formulation by 2^3 Factorial Design**

The objective of the study is to optimize valsartan tablet formulation by 2^3 factorial design for selecting the best combinations of diluent, binder and disintegrant giving fast dissolution of valsartan, a BCS class II drug.

For formulation of valsartan tablets as per 2^3 factorial design the three factors involved are binder, diluent and disintegrant. The two levels of the factor A (binder) are acacia and PVP at 2% concentration each and the two levels of the factor B (disintegrant) are potato starch (15%)and Primojel (5%). The two levels of the factor C (diluent) are lactose and DCP. Eight valsartan tablet formulations each containing 50 mg of valsartan were prepared employing selected combinations of the three factors i.e., binder, disintegrant and diluent as per 2^3 factorial design. The tablets were prepared by wet granulation method.

Much variations were observed in the disintegration and dissolution characteristics of the valsartan tablets prepared employing various combinations of binder (factor A), disintegrant (factor B) and diluent (factor C) as per 2^3 factorial design. Valsartan tablets formulated employing lactose as diluent (F1, Fa, Fb, Fab) disintegrated rapidly within 1 min whereas tablets formulated with DCP disintegrated relatively slowly in 3 – 5 min. Tablets formulated employing lactose as diluent gave higher dissolution rates (K1) and DE30 values when compared to the tablets formulated employing DCP. Formulation Fab (tablets prepared employing lactose, PVP and Primojel), F1 (tablets prepared employing lactose, acacia and potato starch) and Fabc (tablets prepared employing DCP, PVP and Primojel) gave higher dissolution rates and DE30 values and fulfilled the official (I.P 2010) dissolution rate test specification of valsartan tablets .Hence combinations of (i) lactose, PVP , Primojel, (ii) lactose, acacia, potato starch and(iii) DCP, PVP, Primojel are the best combinations of diluent, binder and disintegrant recommended for formulation of valsartan tablets giving rapid and higher dissolution of valsartan, a BCS class II drug.

Case Study 4: **Optimization of Telmisartan Tablet Formulation By 2^3 Factorial Design Employing BCD, Primojel And Tween 80 (IJPSR, 2017; Vol. 8(3): 1178-1183)**

The objective of the study is to optimize the telmisartan tablet formulation employing βCD, Primojel and Tween 80 by 2^3 factorial design to achieve

NLT 85% dissolution in 10 min. For optimization of telmisartan tablets as per 2^3 factorial design the βCD, Primojel and Tween 80 are considered as the three factors. The two levels of the factor A (βCD) are 1:1 and 1:6 ratio of drug: βCD, the two levels of the factor B (Primojel) are 2% and 30% of drug content and the two levels of factor C (Tween 80) are 0 and 2% of drug content. Eight telmisartan tablet formulations employing selected combinations of the three factors *i.e.* βCD, Primojel, and Tween 80 as per 2^3 factorial design were prepared as per the formulae given in Table 10.5.

Table 10.5 Fomulae of Telmisartan tablets prepared employing β-cd, Primojel and Tween 80 as per 2^3 factorial design.

Ingredient	Formulation Code								
(mg/ tablet)	F1	F a	F b	F ab	F c	F ac	F bc	F abc	F opt1
Telmisartan	40	40	40	40	40	40	40	40	40
β-cylodextrin	40	240	40	240	40	240	40	240	140
Primojel	1.6	5.6	24	84	1.6	5.6	24	84	50.2
Tween 80	-	-	-	-	1.6	5.6	1.6	5.6	1.8
Talc	1.6	5.7	2	7.2	1.6	5.8	2.1	7.3	4.6
Magnesium stearate	1.6	5.7	2	7.2	1.6	5.8	2.1	7.3	4.6
Total Weight (mg)	84.8	297	108	378.4	86.4	302.8	109.8	384.2	241.2

F opt 1: Optimised Formulation to achieve NLT 85% Dissolution in 10 Minutes

The tablets were prepared by direct compression method and were evaluated for drug content, hardness, friability, disintegration time and dissolution rate characteristics. The dissolution parameters of the tablets are shown in Table 10.6.

Table 10.6 Dissolution parameters of telmisartan tablets prepared employing β-cd, Primojel and Tween 80 as per 2^3 factorial design.

Formulation	PD_{10} (%)	T50	DE_{20}	$K_1 \times 10^3$ (min^{-1})
Code	(± S.D)	(min)	(%)	(± S.D)
F 1	8.18 ± 1.06	104.0	9.18	6.66 ± 1.78
F a	43.95 ± 1.23	11.9	41.65	57.9 ± 2.19
F b	96.1 ± 0.24	2.1	84.64	324.6 ± 6.28
F ab	88.08 ± 0.18	3.2	78.34	212.7 ± 1.55
F c	6.71 ± 0.27	101.9	6.90	6.96 ± 0.30
F ac	22.15 ± 1.33	27.7	23.58	25 ± 1.68
F bc	95.81 ± 0.46	2.1	84.46	318.03± 11.33
F abc	81.58 ± 0.74	4.0	69.40	169.26 ± 4.0
F opt	85.85 ±1.25	2.8	75.65	204.5± 1.65

The hardness of the tablets was in the range 4.0 - 5.0 kg/cm^2. Weight loss in the friability test was less than 0.92% in all the cases. Telmisartan content of the tablets prepared was within $100 \pm 3\%$. Much variations were observed in the disintegration and dissolution characteristics of the telmisartan tablets prepared. The disintegration times were in the range 20 to 390 sec. Telmisartan tablet formulations F_b and Fbc disintegrated rapidly with in 20 and 40 sec respectively. However, all the telmisartan tablets prepared fulfilled the official (IP 2010) requirements with regard to drug content, hardness, friability and disintegration time specified for uncoated tablets. Dissolution rate of telmisartan tablets prepared was studied in phosphate buffer of pH 7.5.

Telmisartan tablet formulations F_b and F_{bc} gave very rapid dissolution of telmisartan than others. These tablets (F_b and Fbc) gave 96.1% and 95.8% in 10min respectively. Higher levels of βCD and lower levels of Primojel gave low dissolution of telmisartan tablets. The increasing order of dissolution rate (K_1) observed with various formulations was $F_c < F_1 < F_{ac} < F_a < F_{abc} < F_{ab} < F_{bc} < F_b$.

The polynomial equation describing the relationship between the response, percent drug dissolved in 10 min (Y) and the levels of βCD (X_1), Primojel (X_2) and Tween 80 (X_3) based on the observed results was found to be $Y = 55.327 + 3.613(X_1) + 35.072(X_2) - 9.182(X_1 \ X_2) - 3.757(X_3) - 3.317(X_1 \ X_3) + 2.06(X_2 \ X_3) + 1.765(X_1 \ X_2 \ X_3)$ by multiple regression technique.

Estimation of Optimized Levels of the Factors

Estimation of the optimised levels of three factors based on the above polynomial equation is as follows.

1. Factor A is βCD. The two levels of βCD are 1:1 and 1:6 ratio of drug : βCD.

   The transformed potency of the lower level is denoted as -1 and that of higher level is denoted as +1. The average of these two levels which is equal to 0 is selected(X_1=0). The transformed potency of 0 is equal to the average of the two levels that is average of 1:1 and 1:6. The average of 1:1 and 1:6 is equal to 1:3.5. Hence transformed potency of 0 is equal to 1:3.5 ratio of drug: βCD. The optimized level of βCD is selected as 1:3.5ratio of drug : βCD.

2. Factor C is tween 80. The two levels of tween 80 are 0 and 2% of drug .

   The transformed potency of the lower level is denoted as -1 and that of higher level is denoted as +1. The average of these two levels which is equal to 0 is selected (X_3=0). The transformed potency of

0 is equal to the average of the two levels that is average of 0 and 2%. The average of 0 and 2% is equal to 1. Hence transformed potency of 0 is equal to 1% of drug. The optimized level of tween80 is selected as 1% of drug.

3. To estimate the optimized level of factor B (primojel) the selected levels of factor A (X_1=0) and factor C (X_3=0) are substituted in the above polynomial equation.

When X_1=0 and X_3=0 the above polynomial equation reduces to the following equation to achieve NLT 85% dissolution in 10 min.

$$85 = 55.3 + 35.1X_2$$

Therefore, $X_2 = 0.846$

Hence the transformed potency of primojel (factor B) is 0.846 ,which corresponds to the real potency of 27.84%.

Hence based on the above equation, the formulation of optimized telmisartan tablets with NLT 85% dissolution in 10 min require βCD at 1:3.5 ratio of drug: βCD, Primojel at 27.84% of drug content, and Tween 80 at 1% of drug content.

To verify telmisartan tablets were formulated employing the optimized levels of βCD, Primojel and Tween 80. The optimized telmisartan tablet formulation was prepared by direct compression method and the tablets were evaluated. The hardness of the optimized telmisartan tablets was 5.0 kg/sq.cm. Friability (percent weight loss) was less than 0.85%. Disintegration time of the tablets was 25 sec. The optimized telmisartan tablet formulation gave 85.85% dissolution in 10 min fulfilling the target dissolution requirement The dissolution results also indicated validity of optimization technique employed. Hence formulation of telmisartan tablets with NLT 85% dissolution in 10 min could be optimized by 2^3 factorial design.

Chapter 11

Non-Parametric Tests of Significance

Non-parametric (or Distribution-free) tests are mathematical procedures for statistical testing which make no assumptions about the frequency distributions of the variables being assessed.

Non-parametric tests are often used in place of their parametric counterparts when certain assumptions about the underlying population are questionable. For example, when comparing two independent samples, the Mann-Whitney(a non-parametric test) does not assume that the difference between the two samples is normally distributed, whereas its parametric counterpart, the two sample t-test does assume the condition of their being normally distributed. Many non-parametric methods are applicable for ordered statistics. All tests involving ranked data (i.e., data that can be put in order) or involving ordinal data are non-parametric.

Ordinal Data: A set of data is said to the ordinal if the values or observations belonging to it can be ranked (put in order) or have a rating scale attached. You can count and order, but not measure in ordinal data).

LEARNING OBJECTIVES

- To know
- Non parametric tests, their advantages and applications
- To know the principle and applications of
- Sign test for paired data and One Sample Sign Test,
- Mann-Whitney U Test or a Rank Sum Test,
- Kruskal-Wallis Test,
- One Sample Run Test,
- Rank Correlation Test
- Kolmogorov-Smirnov Test,
- Kendal Test of Concordance,
- Median Test for Two Independent Samples,
- Wilcoxon Signed-Rank Test.

Non-parametric tests may be, and often are, more powerful in detecting population differences when certain assumptions are not satisfied.

Non-parametric models differ from parametric models in that the model structure is not specified intially, but is instead determined from the data. The term non-parametric is not meant to imply that such models completely lack parameters, rather, the number and the nature of parameters is flexible and not fixed in advance. Non-parametric models are, therefore, also called distribution free models.

Non-parametric tests are often referred to as distribution-free tests. These tests have the obvious advantage of not requiring the assumption of normality of the assumption of homogeneity of variance. They compare medians rather than means and as a result, if the data have one or two outliers, their influence is negated.

Many non-parametric tests are based on ranked data. Data are ranked by ordering them from lowest to highest and assigning them, in order, the integer values from I to the sample size. Ties are resolved by assigning tied values. the mean of the ranks they would have received if there were no ties, e.g., 117, 119, 119, 125, 128, becomes 1,2.5,2.5,4,5. (If the two 119s were not tied, they would have been assigned the ranks 2 and 3. The mean of 2 and 3 is 2.5.For large samples many non-parametric techniques can be viewed as the usual normal-theory-based procedures applied to ranks.

Types of Non-Parametric Tests

A large number of non parametric tests exist, but in this chapter we shall discuss the following tests which are better known and widely used ones:

1. *Sign test for paired data and One Sample Sign Test*: In these tests, positive or negative signs are substituted for quantitative values.

2. *Mann-Whitney U Test or a Rank Sum Test*: This test is used to determine whether two independent samples have been drawn from the same population.

3. *Kruskal-Wallis Test*: It is another rank sum test which generalized the analysis of variance to enable us to dispense with the assumption the population are normally distributed

4. *One Sample Run Test*: It is a method to determine the randomness with which the sample items have been selected.

5. *Rank Correlation Test*: It is a method for finding correlation coefficient when the data available is not in numerical form but the information is sufficient to rank the data first, second, third, and so on.

6. *Kolmogorov-Smirnov Test*: It is another method of determining the goodness of fit between an observed and theoretical probability distributions.

7. ***Kendal Test of Concordance.*** This test is applicable to situations where we want to test the significance of more than two sets of rankings of individuals.

8. ***Median Test for Two Independent Samples.***

9. ***Wilcoxon Signed-Rank Test.***

Advantages of Non-Parametric Tests

1. Non-parametric tests are generally simple to understand, quicker and easier to apply when the sample sizes are small.

2. They do not require lengthy and laborious calculations and hence they are less time consuming. If significant results are obtained no further work is necessary.

3. They do not require the assumption that a population is distributed in the shape of a normal curve or another specified shape.

4. Non-parametric tests can be applied to all types of data-qualitative (nominal scaling) data in rank form (ordinary scaling) as well as data that have been measured more precisely (internal or ratio scaling).

5. The chief advantage of non-parametric tests is their inherently greater range of applicability because of milder assumptions.

6. Non-parametric tests make fewer and less stringent assumptions (that are more easily met) than the parametric tests.

7. Non-parametric test make less stringent demands of the data.

8. Non-parametric methods provide an air of objectivity when there is no reliable (universally recognised) underlying scale for the original data and there is some concern that the results of standard parametric techniques would be criticized for their dependence on an artificial metric.

Disadvantages of Non-Parametric Methods

Distribution-free or non-parametric methods also have a number of important disadvantages which in practice often outweigh their advantages.

1. Non-parametric tests sometimes pay for freedom from assumptions by a marked lack of sensitivity.

2. Important effects can be entirely missed by a non-parametric technique because of its lack of power or because it produces confidence intervals that are too wide. To put it another way, some of these techniques are very blunt instruments. It may cause it unwise to substitute one of these techniques either because of computational ease or because the data do not quite satisfy some distributional assumption.

3. The competing distribution-tied procedure may be quiet robust to departures from that assumption.

4. An additional related disadvantage is that distribution-free techniques often disregard the actual scale of measurement and substitute either ranks or even a more gross comparison of relative magnitudes. This is not invariably bad, but is often the basis for the loss of power or sensitivity referred to in the preceding paragraph.

5. The another disadvantage is one that is gradually being corrected by research in statistical methodology. At present, we have available a much larger body of techniques in distribution-free hypothesis testing than in distribution-free estimation. To some extent, this is an inevitable accompanying feature of techniques that disregard or weaken the underlying scale of measurement, because estimation is often tied to that scale, whereas hypothesis testing is concerned with selecting one of the two competing decisions or actions.

6. Non-parametric tests are often wasteful of information and thus less efficient than the standard techniques which they replace.

7. The major disadvantage is that non-parametric tests can never be as powerful as their parametric counter parts, when parametric tests can be used. For example sign test uses only the sign of the observation and discard the data as such.

Uses of Non-Parametric Methods

There are four situations in which the use of a distribution-free or non-parametric technique is indicated:

1. When quick or preliminary data analysis is needed. May be you just want to see roughly how things are going with the data. May be somebody is about to leave for Atlantic City to present a paper tomorrow, and his or her chief thought it would be a nice idea to include some statistics in the paper, and you are feeling unusually tender hearted or browbeaten today. May be your computer is suffering an acute exacerbation of some rare electronic disease. May be your computer programmer is manifesting that chronic, epidemic, occupational disability: temporal disorientation with gross distortion of task-time relationships. May be you are doing research in a "primitive" setting without access to modern computational aids.

2. When the assumptions of a competing distribution-tied or parametric procedure are not satisfied and the consequences of this are either unknown or known to be serious. Notice that violation of assumptions per se is not a sufficient reason to reject one of the

classical procedures. As we have emphasized throughout this book, these procedures are often insensitive to such violations. For example, the t-tests and intervals for normal means are relatively insensitive to non-normality. We think that one of the most common error made today is the under application of classical statistical methods because of the overzealous adherence to the letter of the law with respect to assumptions. Remember that assumptions will rarely be exactly satisfied. It is, therefore, as important to have a grasp of the consequences of violations of assumptions. Remember that assumptions will rarely be exactly satisfied. It is, therefore, as important to have a grasp of the consequences of violations of assumptions. A great deal of theoretical research in statistics today is directed toward investigating the robustness of statistical procedures.

3. When data are only roughly scaled; for example, when only comparative rather than absolute magnitudes are available. In dealing with clinical data, perhaps patients can only be classified as better, unchanged, or worse. Perhaps only ranks, that is, largest, second largest, ………., smallest, are available.

4. When the basic question of interest is distribution-free or non-parametric in nature. For example, do we have a random samples, or are these two samples drawn from populations with identical distributions?

The Sign Test for Paired Data

The sign test is the easiest non-parametric test. Its name comes from the fact that it is based on the direction (or sign for plus or minus) of a pair of observations and not their numerical magnitudes.

In such problems, each pair of sample value is replaced by a plus sign if the difference between the paired observation is positive (that is, if the first value exceeds the second value) and by a minus sign if the difference between the paired observation is negative (that is, if the first value is less than the second value) and it is discarded if the difference is zero.

For applying sign test to solve a problem, we count:

Number of + ve signs (+ signs)

Number of - ve signs (- signs)

Number of 0's (I,e., which cannot be included either as positive or negative)

We take the two hypothesis :

Null Hypothesis : $H_0 : p = 0.5$

Alternative Hypothesis: $H_1 : p \neq 0.5$. It is a case of two tailed test. If you look carefully at the above two hypothesis, you will see that the situation is similar to a fair coin toss. If we toss a fair coin 30 times, then probability 'p' would be 0.5 and we would expect about fifteen heads and fifteen tails. In that case, we would use the binomial distribution as the appropriate sampling distribution. We also know that when np and nq are each atleast 5, we can use the normal distribution to approximate the binomial. Thus, we can apply normal distribution test. We calculate the standard error of the proportion p as given by :

$$\text{Standard error of proportion p : S.E. (p) : } \sigma_\rho - \sqrt{\frac{pq}{n}}$$

The two limits of acceptance region at 0.05 level of significance are:

$$p_{H_0} + 1.96\sigma_p \text{ and } p_{H_0} - 1.96\sigma_p$$

It is important to note that if the conditions that both *np* and *nq* are not greater than 5, then we must use the binomial instead of normal distribution test.

Example 1: Use the sign test to see if there is a difference between the blood gluclose measures before and after administering a particular medicine in a group of subjects. Use the 0.05 significance level.

Before:	33	36	41	32	39	47	34	29	32	34	40	42	33	36	27
After :	35	29	38	34	37	47	36	32	30	34	41	38	37	35	28

Solution: Table for before and after treatment

Before(1)	:	33	36	41	32	39	47	34	29	32	34	40	42	33	36	27
After(2)	:	35	29	38	34	37	47	36	32	30	34	41	38	37	35	28
Signs [(1) - (2)]	:	-	+	+	-	+	0	-	-	+	0	-	+	-	+	-

If we count the bottom row of above table, we get

Number of '+' signs = 6

Number of '-' signs = 7

Number of 0's = 2

Total sample size =15.

When we use the sign test, we exclude the evaluation O's as we are testing perceivable differences. We notice that we have 6 plus signs and 7 minus signs, for a total of 13 usable responses. If there is no difference between before and after, *p* (the probability that the first figure exceeds the second figure) would be 0.5 and we would expect to get equal number of plus and minus signs. We set up the hypothesis as :

1. *Null Hypothesis*: $H_0: p \neq 0.5 \Rightarrow$ there is no difference between the two concentrations before and after.

$\Rightarrow$ It is a case of two tailed test.

Sample size: $n = 13$.

Proportion: $p = \dfrac{6}{13} = 0.46$

$\therefore \qquad q = 1 - p = 1 - 0.46 = 0.54.$ Also $q = \dfrac{7}{13} = 0.54.$

Here $np = 13 \times \dfrac{6}{13} = 6$, and $nq = 13 \times \dfrac{7}{13} = 7$ are both greater than 5, so we use the normal distribution to approximate the binomial

Level of significance: $\alpha = 0.05$

Z_σ for two tailed test $= 1.96$ (from the tables)

Standard error of the proportion p : S.E. $(p) = \sigma_p = \sqrt{\dfrac{pq}{n}}$

$\therefore \qquad \sigma = \sqrt{\dfrac{0.46 \times 0.54}{13}} = \sqrt{0.2484} = 0.498$

Limits of Acceptance Region.

Upper Limit: $= 0.5 + (1.96) \times (0.498) = 1.476$

Lower Limit: $= 0.5 - (1.96) \times (0.498) = -0.476$

Decision: The two limits of acceptance region are 1.476 and - 0.476. The sample proportion $p = 0.46$ lies within these two limits, so the sample proportion falls within the acceptance region for this hypothesis. We accept the null hypothesis H_0 ⇒ there is no difference between blood gluclose levels before and after medication at 0.5 level of significance.

One Sample Sign Test

One-sample sign test is a non parametric test which is used as an alternative to the one-sample t-test, where the null hypothesis is $\mu = \mu_0$ against a suitable alternative hypothesis. The only assumptions underlying the sign test are: the population sample is continuous and symmetrical. We assume that the population is continuous so that there is zero probability of getting a value equal to μ_0, and we do not even need the assumption of symmetry if we change the null hypothesis to $\mu = \mu_0$, where μ is the population mean or median.

In the sign test, we replace each sample value exceeding μ_0 with a plus sign and each value less than μ_0 with a minus sign. Then we test the null hypotheses is H_0 that X, the number of plus sign is a value of a random variable having Binomial distribution with parameters n (the total number of plus or minus signs) and probability $\theta = \dfrac{1}{2}$ The two-sided alternative hypothesis $H_1 : \mu \neq \mu_0$ thus becomes $\theta \neq \dfrac{1}{2}$ and one-sided alternatives $\mu < \mu_0$ and $\mu > \mu_0$ becomes $\theta = \dfrac{1}{2}$ and $\theta = \dfrac{1}{2}$ respectively. If a sample value equals μ_0, (which can happen when we deal with rounded data even though the population is continuous), we simply discard it.

In order to perform a sign test, when the sample size is very small, we use the Table of Binomial probabilities and when the sample is large we use the normal approximation to the Binomial distribution with

$$\text{Mean} = n\theta, \text{ and variance} = n\theta (1 - \theta)$$

$$\therefore \qquad \text{S.E. } (\theta) = \sqrt{n\theta(1 - \theta)}$$

The test statistic in this case is, $Z = \dfrac{X - n\theta}{n\theta(1 - \theta)}$

with $0 = \dfrac{1}{2}$, and X the number of plus signs.

***Example* 2:** The following are the measurements of breaking strength of a certain kind of fibres in pounds:

| 163 | 165 | 160 | 189 | 161 | 171 | 158 | 151 | 169 | 162 |
| 163 | 139 | 172 | 165 | 148 | 166 | 172 | 163 | 187 | 173 |

Use the sign test to test the null hypothesis $\mu = 160$ against the hypothesis $\mu > 160$ at the 0.05 level of significance.

***Solution*:**

1. Null Hypothesis: $H_0 : \mu = 160$.

   Alternative Hypothesis: $H_1 : \mu > 160 \Rightarrow$ It is a case of one tailed test.

2. Level of significance: $a = 0.05$.

3. Test statistic: $X =$ The observed number of plus signs.

Replacing each value exceeding 160 with a plus sign, each value less than 160 with a minus sign, and discarding the value which equals to 160, we get :

+   +   0   +   +   +   -   -   +   +   +   -   +   +   -   +   +   +   +   +

Here n = the total number of plus and minus signs = 19.

X = the total number of plus signs = 15.

It is a case of Binomial Distribution.

From the Binomial Table, we find

$$P\,(X \geq 15) = 0.0095 \text{ for } 6 = \text{ and } n = 19.$$

[For $\theta = 0.05$, $P(X \geq 15) = P(X= 15) + P(X= 16) + P(X= 17) + P(X= 18) + P(X= 19)$

$$= 0.0074 + 0.0018 + 0.0003 + 0.0000 + 0.0000 = 0.0095]$$

4. *Decision*: Since *P*-value (= 0.0095) is less than 0.05, so the null hypothesis must be rejected and we conclude that the mean breaking strength of given kind of fibre exceeds 160 pounds.

Example **3**: The following data, (in tons) are the amounts of sulfur oxides emitted by a large pharma plant in 40 days

24	15	20	29	19	18	22	25	27	9
17	20	17	6	24	14	15	23	24	26
19	23	28	19	16	22	24	17	20	13
19	10	23	18	31	13	20	17	24	14

Use the sign test to test the null hypothesis $\mu = 21.5$ against the alternative hypothesis $\mu > 21.5$ at the 0.01 level of significance.

Solution:

1. *Setting up the Hypothesis*:

Null Hypothesis: $H_0 : \mu = 21.5$.

Alternative Hypothesis: $H_1 . \mu > 21.5$ ⟶   It is a case of one tailed test.

2. *Test Statistic*:

Replacing each value exceeding 21.5 with a plus sign, each value less than 21.5 with a minus sign, and discarding the one value which equals 21.5, we get:

+   -   -   +   -   -   +   +   +   -   -   -   -   -   -   +   -   -   +   +   +

-   +   +   -   -   +   +   -   -   -   -   -   +   -   +   -   -   -   +   -

X = the number of plus signs = 16.

n = total number of plus and minus signs = 16 + 24 = 40.

As the sample size, $n = 40$ is very large, so we shall use the normal approximation to Binomial distribution.

$$\therefore \qquad \text{\textit{Test} Statistic: } Z = \frac{X - n\theta}{n\theta(1-\theta)}$$

where $0 = \dfrac{1}{2} = 0.5$, n = sample size.

$$\therefore \qquad Z = \frac{16 - 40(0.5)}{\sqrt{40(0.5)(0.5)}} = \frac{-4}{3.16} = -1.26$$

$$\therefore \qquad |Z| = 1.26 \qquad\qquad |\because n = 40, \theta = 0.5|$$

3. *Level of significance*: Here $\alpha = 0.01$

4. *Critical values*: The critical value $|Z_\alpha|$ for $\alpha = 0.01$ for one-tailed test is 2.33.

Decision: Since $|Z| < |Z_\alpha|$ as *1.26* <2.33, so we accept the null hypothesis.

Worked-out Examples

1. Use the sign test to see if there is a difference between the number of days until collection of an account receivable before and after a collection policy. Use 0.05 significance level.

Before : 30 28 34 35 40 42 33 38 34 45 28 27 25 41 36

After : 34 29 33 32 47 43 40 42 37 44 27 33 30 38 36.

[*Ans*: There is no significant difference before and after new collection policy in the accounts receivable.]

2. The following data show employees' rates of defective work before and after a change in the wage incentive plan. Compare the two sets of data given below to sec if the change lowered the defective units produced. Use the 0.10 level of significance.

Before : 87697108658108

After : 6586981075695.

3. Use the sign test on the data given below to determine whether there is a statistical increase in the values produced by treatment *B* over those produced by treatment *A*:

Subject	:	1	2	3	4	5	6	7	8	9	10
Treatment A	:	46	41	37	32	28	43	42	51	28	27
Treatment B	:	52	43	37	32	31	39	44	53	26	31.

Use 0.05 level of significance.

4. When is the sign test used? The scores under two conditions X and Y obtained by the respondents are given below:

   X :   2   16   8   6   4   8

   Y :   7   12   17   5   12   11.

   Apply the sign test and comment on your findings at 0.05 level of significance.

5. A company claims that if its product is added to an automobile's gasoline tank, the distance travelled in kilometres per litre will improve. To test the claim, 15 different automobiles are chosen and the distances with and without the additive arc measured; the results are shown below. Assuming that the driving conditions arc the same, determine whether there is a difference due to the additive at significance levels of *(a)* 0.05 and *(b)* 0.01.

With additive:	*113*	14.1	9.8	12.5	7.8	12.2	14.3	11.7	13.8	16.0	24.8	11.2	12.8	14.0	12.1
Without additive:	15.7	13.6	10.2	12.3	7.4	11.1	13.4	12.0	13.1	15.7	14.4	11.5	12.0	13.6	11.4

   [Ans. *(a)* There is a difference at the 0.05 level of significance

   (b) There is no difference at the 0.01 level of significance.]

6. The following data represents the number of hours that a rechargeable hedge trimmer operates before a recharge is required.

   1.52.2   0.9   1.3   2.0   1.6   1.8   1.5   2.6   1.2   1.7.

   Use the sign test to test thc hypothesis at 0.05 level of significance that this particular trimmer operates, on the average, 1.8 hours before requiring a recharge.

   [**Hint: H$_0$:** $\mu = 1.8$, *H$_1$:* $\mu \neq 1.8$, we have - + - - + - - - - + - -, so, $n = 10$ and $X = 3$. Apply Binomial Table].

   [Ans. Accept the null hypothesis $H_0{:}\mu = 1.8$]

7. On 12 visits to a doctor's clinic, a patient had to wait in minutes as under:

   17;   32;   25;   15;   28;   25:   20;   12;   35;   20;   26;   24.

   before being seen by the doctor. Use the sign test with a = 0.05 to test the doctor's claim that, on the average, his patient do not wait more than 20   minutes   before   being   examined   by   him.
   [Ans. H_0: $\mu = 20$ is accepted]

Rank Sum Tests

Rank sum tests are a whole family of tests. We shall concentrate only on the following two members of this family:

1. Mann-Whitney U Test
2. Kruskal-Wallis H Test.

Mann-Whitney tests is used when there are only two populations whereas Kruskal-Wallis test is employed when more than two populations are involved. The use of these tests will enable us to determine whether independent samples have been drawn from the same population or different populations have the same distribution.

Mann-Whitney U-Test

It is a non-parametric test used to determine whether two independent samples have been drawn from populations with same distribution. This test is also known as U-Test. This test enables us to test the null hypothesis that both population medians are equal (or that the two samples are drawn from a single population). It requires the two samples to the independent samples of observations measured at least at an ordinal level, i.e., we-can at least say, of any two observations, which is greater. This method does not require the assumption that the differences between the two samples are normally distributed.

This method helps us to determine whether the two samples have come from identical populations it is true that the samples have come from the same population, it is reasonable to assume that the medians of ranks assigned to the values of two samples are more or less the same. The alternative hypothesis H_1 would be that the medians of the populations are not equal. In this case, most of the smaller ranks will go to the values of one sample, while most of the higher ranks will go to the other sample. The test involves the calculation of a statistic usually called U, whose distribution under the null hypothesis is known. In case of small samples (i.e., when the sample size is less than 8), the distribution is tabulated but for samples above 8, there is good approximation using the normal distribution.

Step 1. Set the null hypothesis H_0 and alternative hypothesis H_1.

H_0: Both the medians are equal.

H_1: Both the population medians are not equal. $\Rightarrow$ a case of two tailed test

Step 2. Combine all sample values in an array from smallest to the largest, and assign ranks to all these values. If two or more sample values are identical *(i.e.,* there are tie scores), the sample values are each assigned a rank equal to the mean of the ranks that would otherwise be assigned.

Step 3. Find the sum of the ranks for each of the samples. Let us denote these sums by R_1 and R_2

Also n_1 and n_2 are their respective sample sizes. For convenience, choose n_1 as the smaller size if they are unequal so that $n_1 < n_2$.

A significant difference between the rank sums R_1, and R_2, implies a significant difference between the samples.

Step 4. Calculation of U-Statistic to test the difference between the rank sums.

U Statistic $\quad U = n_1 n_2 + \dfrac{n_1(n_1 + 1)}{2} - R_1$ [Corresponding to Sample 1]

or $\quad U = n_1 n_2 + \dfrac{n_2(n_2 + 1)}{2} - R_2$ [Corresponding to Sample 2]

The sampling distribution of U is symmetrical and has a mean and variance given by the formulae:

Mean: $\quad = \mu_U \dfrac{n_1 n_2}{2}$

Variance : $\quad = \dfrac{n_1 n_2 (n_1 + n_2 + 1)}{2}$

If n_1 and n_2 are both atleast equal to 8, it turns out that the distribution of U is nearly normal and one could use the statistic Z, where

$$Z = \dfrac{U - \mu_U}{S.E.(U)}$$

is normally distributed with mean 0 and variance 1.

Step 5. Level of significance: Take $\alpha = 0.05$

Step 6. Critical Region: Accept H_0 if $|Z| < |Z_\alpha|$, where Z_α is the tabled value of Z for the given level of significance . The values of Z_α are given by the normal table.

Remarks: In many applications Mann-Whitney test is used in place of two samples, t-test when normality assumption is questionable.

This test can be applied when the observations in a sample of data are ranks that is ordinal data rather than direct measurements.

***Example* 4:** The total alkaloid contents of two samples of a medicinal plant, measured in mg/kg, are found to be as follows:

Sample A:	2.1	4.0	6.3	5.4	4.8	3.1	6.1	3.3		
Sample B :	4.1	0.6	3.1	2.5	4.0	6.2	1.6	2.2	1.9	3.4

Test the hypothesis, at the 0.05 level of significance, that the average alkaloid contents of the two samples are equal against the alternative that they are unequal.

Solution: Setting up of Hypothesis

Null Hypothesis: $H_0 : \mu_1 = \mu_2$, *i.e.,* the average alkaloid contents of the two samples are equal.

Alternative Hypothesis: $H_1 : \mu_1 \neq \mu_2$, *i.e.,* the average alkaloid contents of the two samples are not equal.

$\Rightarrow$ It is a case of two tailed test.

2. Level of significance: Here $\alpha = 0.05$.

3. Computation of U-statistic:

The observations of both the samples are arranged in ascending order and ranks from 1 to 18 are assigned.

Original Data :	0.6	1.6	1.9	2.1	2.2	2.5	3.1	3.3	3.7	4.0	4.0	4.1	4.8	5.4	5.4	6.1	6.2	6.3
Rank	1	2	3	4	5	6	7	8	9	10.5	10.5	12	13	14.5	14.5	16	17	18

The ranks of the observations belonging to the smaller samples are underscored (put in bold form)

$$R_1 = 4 + 8 + 9 + 10.5 + 13 + 14.5 + 16 + 18 = 93$$

$$R_2 = 1 + 2 + 3 + 5 + 6 + 7 + 10.5 + 12 + 14.5 + 17 = 78.$$

Also, $n_1 = 8$; $n_2 = 10$. For the calculation of U-statistic we take $R_1 =$ sum of the ranks of smaller groups.

$\therefore$ **U-Statistic:** $U = n_1 n_2 + \dfrac{n_1(n_1 + 1)}{2} - R_1$

$$= 8 \times 10 + \dfrac{8 \times 9}{2} - 93 = 80 + 36 - 93 = 23.\,2$$

$\therefore$ **Mean of U** $= \mu_U = \dfrac{n_1 n_2}{2} = \dfrac{10 \times 8}{2} = 40$

Variance of U $= \sigma_U^2 = \dfrac{(n_1 \times n_2)(n_1 + n_2 + 1)}{2}$

$$= \dfrac{10 \times 8(10 + 8 + 1)}{12} = \dfrac{380}{3} = 126.1 = 126.67$$

$$\sigma_U = \sqrt{126.67} = 11.25$$

Here $n_2 = 10$, so we can use the statistic

$$Z \,, \; Z = \frac{U - \mu_U}{\sigma_U} = \frac{23 - 40}{11.25} = \frac{-17}{11.25} = -1.51$$

$$|Z| = 1.51$$

The tabled value of Z_α at $\alpha = 0.05$ is 1.96.

Now, $|Z| < |Z_\alpha|$ as $1.51 < 1.96$, so we accept the null hypothesis H_0 and conclude that there is no significant difference in the average alkaloid contents of the two samples.

***Example* 5:** The following are the weight gains (in pounds) of two random samples of childern fed on two different diets but otherwise kept under identical conditions:

Diet I:	16.3	10.1	10.7	13.5	14.9	11.8	14.3	10.2
	12.0	14.7	23.6	15.1	14.5	18.4	13.2	14.0
Diet II:	21.3	23.8	15.4	19.6	12.0	13.9	18.8	19.2
	15.3	20.1	14.8	18.9	20.7	21.1	15.8	16.2.

Use U test at 0.01 level of significance to test the null hypothesis that the two population samples are identical against the alternative hypothesis that on the average the second diet produces a greater gain in weight.

***Solution* :**

1. Setting up of Hypothesis:

   Null Hypothesis: $H_0 : \mu_1 = \mu_2$

   Alternative Hypothesis: $H_1 : \mu_1 < \mu_2$ It is a case of one tailed test.

2. Level of significance: $\alpha = 0.01$

3. Test Statistic: Ranking the data jointly according to size we find the values of first sample occupy the ranks: 21, 1, 3, 8, 15, 4, 11, 2, 5.5, 13, 31, 16, 12, 22, 7 and 10 (the fifth and sixth values are both 12, so we assigned each the rank 5.5).

   $\therefore R_1 =$ The sum of the ranks of the first sample

$$= \; 21 + 1 + 3 + 8 + 15 + 4 + 11 + 2 + 5.5 + 13 + 31 + 16 + 12 + 22 + 7 + 10$$

$$= \; 181.5.$$

Also $n_1 = 16$, $n_2 = 16$.

$\therefore \qquad$ Statistics $U = n_1 n_2 + \dfrac{n_1(n_1 + 1)}{2} - R_1$

$$U = 16 \times 16 \left[\frac{16(16 + 1)}{2} \right] - 181.5 = 210.5$$

As the sample sizes n_1 and n_2 are both greater than 8, so the distribution of U is nearly normal with

$$\text{Mean}: \mu_U = \frac{n_1 n_2}{2} = \frac{16 \times 16}{2} = 128$$

$$\text{Variance}: \sigma_U^2 = \frac{(n_1 \times n_2)(n_1 + n_2 + 1)}{12}$$

$$= (16 \times 16)(16 + 16 + 1)/12 = 704$$

$$\therefore \quad \text{S.E. (U)} = \sigma_u = \sqrt{704} = 26.53$$

$$\text{Test statistics}: \quad Z = \frac{U - \mu_u}{\text{S.E.(U)}} = \frac{210.5 - 128}{26.53} = \frac{82.5}{26.53} = 3.11$$

Critical value: From the normal tables the value of Z_α for $\alpha = 0.01$ is 2.33.

$$\therefore \quad \left| z_\alpha \right| = 2.33.$$

4. *Result*: Reject the hypothesis if the calculated value of $|Z|$ is more than the tabled value of $|Z_\alpha|$

Here $|z| > |z_\alpha|$ as 3.11 > 2.33, so the null hypothesis is rejected and we conclude that on the average the second diet produces a greater gain in weight.

Kruskal-Wallis Test or H-Test

The Kruskal-Wallis test is a generalization of Mann-Whitney U-test to the case of $k > 2$ samples. This test is also known as Kruskal-Wallis H-test. It is used to test the null hypothes is H_0 that k independent samples are drawn from the identical population. This test is an alternative non-parametric test to the F-test for testing, the equality of means in the one factor analysis of variance when the experimenter wishes to avoid the assumption that the samples were selected from the normal populations. Let n_i. $(i = 1, 2, 3......, k)$ be the number of observations in the i th sample. First we combine all k samples and arrange them to get $n = n_1 + n_2 + n_3 + ... + n_k$ observations in ascending order, substituting the appropriate rank from 1, 2, ... ,n for each observation. In the case of ties (identical observations), we follow the usual procedure of replacing the observations by the means of the ranks that the observations would have if they were distinguishable. The sum of the ranks corresponding to the n observations in the i th sample is denoted by the random variable R_i. Now let us consider the statistic.

$$H = \frac{12}{n(n+1)} \sum_{i-1}^{k} \frac{R_i^2}{n_i} - 3(n+1)$$

which is approximated very well by a chi-square distribution with *k*-1 degrees of freedom when H_0 is true and if each sample consists of at least five observations.

If *H* falls in the critical region $H > \Psi_\alpha^2$ with $v = k - 1$ degrees of freedom, we reject H_0 at α level of significance; otherwise, we accept *H*.

***Example* 6:** Three analytical methods I, II and III are used to estimate the drug content in a batch of tablets. The following are the results obtained. Determine at $\alpha = 0.05$ significance level whether there is a difference between the analytical methods.

Method I :	78	62	71	58	73
Method II :	76	85	77	90	87
Method III :	74	79	60	75	80

***Solution*:** *Null Hypothesis*: H_0: There is difference between the analytical methods.

Alternative Hypothesis: H_1: There is no difference between the three analytical methods.

Since there are three methods (I, II and III) and each group consists of 5 estimates, so we have $N_1 = N_2 = N_3 = 5$, *and* $N = N_1 + N_2 + N_3 = 5 + 5 + 5 = 15$. Arranging all these estimates in increasing order of magnitude and assigning appropriate ranks, we get:

Estimates:	58	60	62	61	73	74	75	76	77	78	79	80	85	87	90
Ranks:	1	2	3	4	5	6	7	8	9	10	11	12	13	14	15

Rank Table

Method 1	10	3	4	1	5	$23 = R_1$
Method 2	8	13	9	15	14	$59 = R_2$
Method 3	6	11	2	7	12	$38 = R_3$

$$\therefore \quad H = \frac{12}{N(N+1)} \left[\frac{R_1^2}{N_1} + \frac{R_2^2}{N_2} + \frac{R_3^2}{N_3} \right] - 3(N+1)$$

$$\frac{12}{15 \times 16} \left[\frac{(23)^2}{5} + \frac{(59)^2}{5} + \frac{(38)^2}{5} \right] - 3 \times 16$$

$$H = \frac{1}{100}[529 + 3481 + 1444] - 48 = \frac{5454}{100} - 48 = 6.54$$

Here degrees of freedom = k - 1 = 3 – 1 = 2.

Also level of significance : $\alpha = 0.05$

$\therefore \Psi^2$ (for 2 d.f. and $\alpha = 0.05$) = $\Psi_{0.05,2}$ = 5.991. [From Ψ^2-tables]

Decision: Reject H_0 if $H > \Psi^2_{0.05}$

Now 6.54 > 5.991 $\Rightarrow H > \Psi^2_{0.05}$

the null hypothesis H_0 is rejected and the alternative hypothesis H_1 is accepted. We conclude that there is no difference in analytical methods.

***Example* 7**: Four paracetmol tablet brands were tested for dissolution rate and the data (mg dissolved in 30 min) are given below. Test at 0.05 significance level whether there is a product (Brand) difference in dissolution rate.

Brand 1	72	80	83	75	
Brand 2	81	74	77		
Brand 3	88	82	90	87	80
Brand 4	90	71	77	70	

Solution: Null Hypothesis: H_0: There is a significant difference between the brands.

Alternative Hypothesis: H_1: There is no difference between the brands.

$\therefore \qquad n = n_1 + n_2 + n_3 + n_4 = 4 + 3 + 5 + 4 = 16.$

Arranging these data in increasing order of magnitude and assigning appropriate ranks, we have

Table of Ranks

Data:	70	71	72	74	75	77	77	80	80	81	82	83	87	88	90	90
Ranks:	1	2	3	4	5	6.5	6.5	8.5	8.5	10	11	12	13	14	15.5	15.5

Sum of Ranks

Brand 1	3	8.5	12	5		28.5	$= R_1$
Brand 2	10	4	6.5			20.5	$= R_2$
Brand 3	14	11	15.5	13	8.5	62	$= R_3$
Brand 4	15.5	2	6.5	1		25	$= R_4$

Here $N_1, = 4,$ $\qquad N_2, = 3$ $\qquad N_3, = 5$, $N_4, = 4$

$\therefore \qquad N = N_1 + N_2 + N_3 + N_4 = 4 + 3 + 5 + 4 = 16.$

Also $R_1 = 28.5,\ R_2 = 20.5,\ R_3 = 62,\ R_4 = 25.$

$$\therefore \qquad H = \frac{12}{N(N+1)}\left[\frac{R_1^2}{N_1} + \frac{R_2^2}{N_2} + \frac{R_3^2}{N_3}\right] - 3(N+1)$$

$$H = \frac{12}{16\times17}\left[\frac{(28.5)^2}{4} + \frac{(20.5)^2}{3} + \frac{(62)^2}{5} + \frac{(25)^2}{4}\right] - 3\times17$$

$$\frac{12}{16\times17}\left[\frac{812.25}{4} + \frac{420.25}{3} + \frac{3844}{5} + \frac{625}{4}\right] - 51$$

$$= \frac{12}{16\times17}\left[203.06 + 140.08 + 768.80 + 156.25\right] - 51$$

$$= \frac{12}{16\times17}(1268.19) - 51 = 55.95 - 51 = 4.95$$

Degrees of freedom = k-1 = 4 -1=3

Level of significance: Here $\alpha = 0.05$

The value of Ψ^2 for 3 degrees of freedom for $\alpha = 0.05$ is 7.81 [From Ψ^2 table]

Decision: Since the value of H is less than the tabled value of Ψ^2 at a = 0.05 for 3 degrees of freedom as 4.95 < 7.81, so we accept the null hypothesis H_0, there is a significant difference between the brands with respect to dissolution rate.

Example 8: The following data refers to the amount of drug excreted (mg) in urine in 24hrs following the administration of three drug products A, B, C.

Product A: 94, 88, 91, 74, 87, 97

Product B: 85, 82, 79, 84, 61, 72, 80

Product C: 89, 67, 72, 76, 69.

Use the H-test at the 0.05 level of significance to test the null hypothesis that the three drug products are equally bioavailable.

Solution:

1. Setting up of Hypothesis:

    Null Hypothesis: H_0: $\mu_1, = \mu_2 = \mu_3$.

    Alternative Hypothesis H_1: μ_1, μ_2, μ_3 are not all equal.

2. Level of significance: Here $\alpha = 0.05$.

3. Calculation of Test Statistic: Ranking the data jointly from 1 to 18, we find that

$$R_1 = 6+13 + 14 + 16 + 17+18 = 84$$

$$R_2 = 1 + 4.5 + 8 + 9 + 10 + 12 = 55.5$$

$$R_3 = 2 + 3 + 4.5 + 7 + 15 = 31.5$$

[∴ there is only one tie and the tied values are each assigned the rank 4.5]

Also $\quad N_1, = 6, \qquad N_2 = 7, \qquad N_3 = 5$

∴ $\qquad N = N_1 + N_2 + N_3 = 6 + 7 + 5 = 18.$

∴ $\qquad$ Test statistic: $H = \dfrac{12}{N(N+1)}\left[\dfrac{R_1^2}{N_1} + \dfrac{R_2^2}{N_2} + \dfrac{R_3^2}{N_3}\right] - 3(N+1)$

Or $\qquad H = \dfrac{12}{16 \times 17}\left[\dfrac{(84)^2}{6} + \dfrac{(55.5)^2}{7} + \dfrac{(31.5)^2}{5}\right] - 3 \times 19$

$$= \dfrac{2}{17}\left[1176 + \dfrac{3080.25}{7} + \dfrac{992.25}{5}\right]$$

$$\dfrac{2}{57}\,[1176 + 440.04 + 198.45] - 57$$

$$= \dfrac{2}{57} \times 1814.49 - 57 = 63.67 - 57 = 6.67$$

Degrees of Freedom = k -1 = 3-1 = 2

∴ Ψ^2 (for 2 d.f. and α = 0.05) = 5.991. [From Ψ^2 table]

Decision: Since *H* = 6.67 exceeds = 5.991, so the null hypothesis is rejected. We accept the alternative hypothesis H_1, and conclude that the three drug products are not equally bioavailable.

Practice Exercise

1. What is Mann-Whitney U test? When is it used?

2. Explain the Mann-Whitney U test with the help of an example.

3. A farmer wishes to determine whether there is a difference in yields between two different varieties of wheat, I and II. The following data shows the production of wheat per unit area using the two varieties. Can the farmer conclude at signficance levels of *(a)* 0.05 and *(b)* 0.01 that a difference exists?

Wheat I	15.9	15.3	16.4	14.9	15.3	16.0	14.6	15.3	14.5	16.6	16.0
Wheat II	16.4	16.8	17.1	16.9	18.0	15.6	18.1	17.2	15.4		

4. Instructors A and B both teach a first course in chemistry at XYZ University. On a common final examination, their students received the marks shown as given below. Test at the 0.05 significance level the hypothesis that there is no difference between the two instructors' grades.

A	88 75	92	71	63	84	55	64	82	96					
B		72	65	84	53	76	80	51	60	57	85	94	87	73	61

5. A company wishes to determine whether there is a difference between two brands of gasoline, A and B. The following data shows the distances travelled in kilometres per litre for each brand. Can we conclude at the 0.05 significance level *(a)* that there is a difference between the brands and *(b)* that brand B is better than brand $A?$

A	15.2	14.4	14.6	16.3	15.9	14.7	15.4	15.6	15.3	16.0
B	16.7	14.9	15.0	15.8	16.9	15.5	15.7	14.8	16.4	16.5

6. Explain Kruskal-Wallis test with the help of an example.

7. In an experiment to determine which of three different missile systems is preferable, the propellant burning rates was measured. The data after coding is given below.

System	I	24.0	16.	7	22.8	19.8	18	.9				
System	II	23.2	19.	8	18.1	17.6	20	.2	17	.8		
System	III	18.4	19.	1	17.3	17.3	19	.7	18	.9	18.8	19.3

Use the Kruskal-Wallis test and a significance level of a = 0.05 to test the hypothesis that propellant burning rates are the same for the three missiles.

Hint. Let $H_0 : \mu_1 = \mu_2 = \mu_3$

Table Ranks for Propellant Burning Rates

Missile System I	Missile System II	Missile System III
19	18	7
1	14.5	11
17	6	2.5
14.5	4	2.5
9.5	16	13
	5	9.5
		8
		12
$R_1 = 61.0$	$R_2 = 63.5$	$R_3 = 65.5$

Also n1 = 5, $n_2 = 6$, $n_3 = 8$ and n = $n_1 + n_2 + n_3 = 19$

$$\therefore \qquad H = \frac{12}{n(n+1)}\left[\frac{R_1^2}{n_1} + \frac{R_2^2}{n_2} + \frac{R_3^2}{n_3}\right] - 3(n+1)$$

$$\frac{12}{19 \times 20}\left[\frac{(61.0)^2}{5} + \frac{(63.5)^2}{6} + \frac{(65.5)^2}{8}\right] - 3 \times 20 = 1.66$$

Also, the value of for 18 degrees of freedom at $\alpha = 0.05$ is 28.869.[from Ψ^2 tables] Since $H <$as $1.66 < 5.991$, so we accept the null hypothesis H_0]

8. The same mathematics papers were marked by three teachers *A, B* and *C*. The final marks were recorded as follows :

Teacher A	73	89	82	43	80	73	66	60	45	93	36	77
Teacher B	88	78	48	91	51	85	74	77	31	78	62	76
Teacher C	68	79	56	91	71	71	87	41	59	68	53	79

Use Kruskal-Wallis test, at the 0.05 level of significance to determine if the marks distributions given by the 3 teachers differ significantly.

9. The following data represent the operating times in hours for 3 types of scientific pocket calculators before a recharge is required.

Calculator

A	B	C
4.9	5.5	6.4
6.1	5.4	6.8
4.3	6.2	5.6
4.6	5.8	6.5
5.3	5.5	6.3
5.2	66	
4.8		

Use Kruskal-Wallis test, at the 0.01 level of significance, to test the hypothesis that the operating times for all three calculators are equal.

10. The following are the miles per gallon which a test driver got for 10 tankfuls of each of three kinds of gasoline:

Gasoline A :	20	31	24	33	23	24	28	16	19	26
Gasoline B :	29	18	29	19	20	21	34	33	30	23
Gasoline C :	19	31	16	26	31	33	28	28	25	30

Use the Kruskal-Wallis test at the 0.05 level of significance to test whether or not there is a difference in the actual average mileage yield of the three kinds of gasoline.

11. Three groups of guinea pigs were injected, respectively, with 0.5, 1.0 and 1.5 milligrams of a tranquilize and the following are the numbers of seconds it took them to fall asleep:

0.5-mg dose :	8.2	10.0	10.2	13.7	14.0	7.8	12.7	10.9		
1.0-mg dose :	9.7	13.1	11.0	7.5	13.3	12.5	8.8	12.9	7.9	10.5
1.5-mg dose :	12.0	7.2	8.0	9.4	11.3	9.0	11.5	8.5.		

Use the *H* test at the 0.01 level of significance to test the null hypothesis that the difference in dosage have no effect on the length of time it takes guinea pigs to fall asleep.

12. To compare four bowling balls, a professional bowler bowls five games with each ball and gets the following results:

Ball A:	208	220	247	192	229
BallB:	216	196	189	205	210
Ball C :	226	218	252	225	202
BalfD:	212	198	207	232	221.

Use the Kruskal-Wallis test at the 0.05 level of significance to test whether or not the bowler can expect to score equally well with the four bowling balls.

Answers

1. *(a)* Yes   *(b)* Yes.

4. There is no difference at 0.05 level of significance.

5. *(a)* Yes   *(b)* Yes.

7. H_0: The propellant burning rates are the same for the three missiles is accepted.

8. The marks given by the three teachers differ significantly.

9. The times for the three calculators are not equal.

10. The null hypothesis cannot be rejected; $H = 0.86$.

11. The null hypothesis is accepted.

12. The null hypothesis is accepted.

 ## Spearman's Rank Correlation Test

Spearman's Rank Correlation

Spearman rank correlation is a measure of the correlation that exists between the two sets of ranks, or it is a measure of degree of association between the variables that we would not have been able to calculate otherwise.

Rank Correlation Coefficient

The simple correlation coefficient (r) measures the linear relationship between two variables X and Y. If ranks 1, 2,..., n are assigned to the X observations in order of magnitude and similarly to the F observations, and if these ranks are then substituted for the actual numerical values into the formula for r, we obtain the non-parametric counterpart of the conventional correlation coefficient. A correlation coefficient calculated in this manner is known as the Spearman's rank correlation coefficient and is denoted by r_s. When there are no ties among either set of measurements, the formula for r_s reduces to a much simpler expression, which is given below.

Rank Correlation Coefficient. A non-parametric measure of association between two variables X and Y is given by the rank correlation coefficient

$$r_s = 1 - \frac{6\sum_{i-1}^{n} d_i^2}{n(n^2 - 1)} \qquad .....(A)$$

Where $\qquad$ Σ = notation meaning "the sum of"

$\qquad$ d_i = the difference between the ranks assigned to x_i and y_i

$\qquad$ n = the number of paired observations

$\qquad$ r_s = coefficient of rank correlation

The value of r_s will usually be close to the value obtained by finding r based on numerical measurements and is interpreted in much the same way. As before, the values of r_s will range from - 1 to + 1. A value of + 1 or - 1 indicates perfect association between X and Y, the plus sign occurring for identical rankings and the minus sign occurring for reverse rankings. When r_s is close to zero, we would conclude that the variables are uncorrelated.

Advantages of Rank Correlation Coefficient

1. We do not assume the underlying relationship between X and Y to be linear and, therefore, when the data possess a distinct curvilinear relationship, the rank correlation coefficient will likely be more reliable than the conventional measure.

2. A second advantage in using the rank correlation coefficient is the fact that no assumptions of normality are made concerning the distributions of X and Y.

3. The greatest advantage occurs when one is unable to make meaningful numerical measurements but nevertheless can establish

rankings. Such is the case, for example, when different judges rank a group of individuals according to some attribute. The rank correlation coefficient can be used in this situation as a measure of the consistency of the two judges.

***Example* 9:** The following data refers to the *invitro* release rate and *invivo* c_{max} observed following the testing of 10 brands of a telmesartan tablets.

Brand	Dissolution rate(mg/30min)	C_{max} (µg/ml)
1	14	0.9
2	16	1.1
3	28	1.6
4	17	1.3
5	15	1.0
6	13	0.8
7	24	1.5
8	25	1.4
9	18	1.2
10	31	2.0

Calculate the rank correlation coefficient to measure the degree of relationship betweenthe dissolution rate and C_{max}.

***Solution*:** Let X and Y respectively represent the dissolution rate and C_{max}. We assign ranks to each set of measurements with the rank of 1 assigned to the lowest number in each set, the rank of 2 to the second lowest number in each set, and so forth, until the rank 10 is assigned to the largest number. The following table shows the individual rankings of the measurements and the differences in ranks for the 10 pairs of observations

Table 11.1 Rankings for dissolution rate and C_{max}

Brand	x_i	y_i	$d_i = (x_i - y_i)$	d_i^2
1	2	2	0	0
2	4	4	0	0
3	9	9	0	0
4	5	6	-1	1
5	3	3	0	0
6	1	1	0	0
7	7	8	- 1	1
8	8	7	1	1
9	6	5	1	1
Players	10	10	0	0
				$\sum d_i^2 = 4$

Substituting into the formula for r_s, we get

$$r_s = 1 - \frac{6\sum_{i-1}^{n} d_i^2}{n(n^2-1)} = 1 - \frac{(6)(4)}{(10)(100-1)} = 1 - 0.0242 = 0.9758$$

Hence, r_s= 0.9758

Also r_s = 0.9758 indicates a high positive correlation between the dissolution rate and C_{max} of drug products.

***Example* 10:** Ten brands of ibuprofen tablets are tested for hardness, disintegration time and dissolution rate. Based on the results ranks are assigned to the brands for each character as follows.

Hardness (H) (Low H = Rank 1)	1	5	4	8	9	6	10	7	3	2
DT (Low DT = Rank 1)	4	8	7	6	5	9	10	3	2	1
Dissolution rate(DR) (High DR = 1)	6	7	8	1	5	10	9	2	3	4

Use the rank correlation coefficient to discuss the relationship between the three characters pair wise.

H (R_1)	DT (R_2)	DR (R_3)	$D_1 = R_1 - R_2$	D_1^2	$D_2 = R_1 - R_3$	D_2^2	$D_3 = R_2 - R_3$	D_3^2
1	4	6	-3	9.	-5	25	-2	4
5	8	7	-3	9	-2	4	1	1
4	7	8	-3	9	-4	16	-1	1
8	6	1	2	4	7	49	5	25
9	5	5	4	16	4	16	0	0
6	9	10	-3	9	-4	16	-1	1
10	10	9	0	0	1	1	1	1
7	3	2	4	16	5	25	1	1
3	2	3	1	1	0	0	-1	1
2	1	4	1	1	-2	4	-3	9
				$\sum D_1^2 = 74$		$\sum D_2^2 = 156$		$\sum D_3^2 = 44$

$$r_{12}\ (\text{H\&DT}) = 1 - \frac{6\sum D_1^2}{N^3 - N} = 1 - \frac{6 \times 74}{990} = 1 - 0.45 = 0.55$$

$$r_{13}\ (\text{H\&DR}) = 1 - \frac{6\sum D_2^2}{N^3 - N} = 1 - \frac{6 \times 156}{990} = 1 - 0.945 = 0.55$$

$$r_{23}\ (\text{DT\&DR}) = 1 - \frac{6\sum D_3^2}{N^3 - N} = 1 - \frac{6 \times 44}{990} = 1 - 0.27 = 0.73$$

Since rank correlation is the maximum between DT and DR , we can say that there is a good relationship between disintegration time and dissolution rate of tablets. A moderate relationship exit between hardness and disintegtation time and hardness and dissolution rate.

Kolmogorov-Smirnov Test

The statisticians A.N. Kolmogorov and N.V. Smirnov developed the Kolmogorov-Smirnov test. It is a simple non-parametric test for testing whether there is a significant difference between an observed frequency distribution and a theoretical frequency distribution. In short, this test is known as K-S test. The K-S test is, therefore, another measure of the goodness of fit of a frequency distribution, as was the chi-square test. Its basic advantages are:

1. It is a more powerful test.

2. It is easier to use, since it does not require that the data be grouped in any way.

K-S Statistic

The K-S statistic is the maximum absolute deviation of expected relative frequency F_e and the observed relative frequency F_0. It is denoted by D_n.

$$\text{K-S statistic:} \qquad D_n = \max [F_e - F_0]$$

The K-S test is always a one tailed test for a given the level of significance α.

The critical values for D_n can be tabulated by using the table. We compare the calculated value of D with the critical value of D from the table.

If the tabled value for the chosen significance level is greater than the calculated value of D_n, then we will accept the null hypothesis H_0.

***Example* 11:** Below is the table of observed frequencies, along with the frequency to the observed under a normal distribution.

(a) Calculate the K-S statistic.

(b) Can we conclude that this distribution does in fact follow a normal distribution? Use 0.01 level of significance.

Test Score	51-60	61-70	71-80	81-90	91-100
Observed Frequency	30	100	440	500	130
Expected Frequency	40	170	500	390	100

***Solution*:** Null Hypothesis : H_0 : The observed distribution follows a normal distribution or observed frequency is equal to expected frequency

Table 11.2 Calculation of K-S Statistic.

Test Score	Observed Frequency	Observed Cumulative Frequency	Observed Relative Frequency (F_0)	Expected Frequency	Expected Cumulative Frequency	Expected Relative Frequency (F_e)	D - $\lvert F_e - F_0 \rvert$
51-60	30	30	$\frac{30}{1200} = 0.025$	40	40	$\frac{40}{1200} = 0.0.33$	0.008
61-70	100	130	$\frac{130}{1200} = 0.108$	170	210	$\frac{210}{1200} = 0.175$	- 0.067
71-80	440	570	$\frac{570}{1200} = 0.475$	500	710	$\frac{710}{1200} = 0.592$	0.117
81-90	500	1070	$\frac{1070}{1200} = 0.89$	390	1100	$\frac{1100}{1200} = 0.92$	0.029
91-100	130	1200	$\frac{1200}{1200} = 1$	100	1200	$\frac{1200}{1200} = 1$	0

(a) $\therefore$ K-S statistic: $D_n = \max \lvert F_e - F_0 \rvert = 0.117$

(b) The tabulated value of D_n for $n = 5$ and $\alpha = 0.01$ is 0.510.

Since, the table value of D_n (= 0.510) is greater than the calculated value of D_n (= 0.117), we accept the null hypothesis H_0, the observed distribution follows a normal distribution.

Exercise **12:**

1. The following table shows how 10 students, arranged in alphabetical order, were ranked according to their achievements in both the laboratory and lecture sections of a computer course. Find the coefficient of rank correlation.

Laboratory :	8	3	9	2	7	10	4	6	1	5
Lecture :	9	5	10	1	8	7	3	4	2	6

Difference of ranks (D)	-1	- 2	-1	1	-1	3	1	2	-1	-1
D^2		1	4	1	1	1	9	1	4	1	1
					$\Sigma D^2 = 24$					

$$\therefore \qquad r_s = 1 - \frac{6\sum D^2}{N(N^2 - 1)} = 1 - \frac{6(24)}{10(10^2 - 1)} = 0.8545$$

2. In a contest, two judges were asked to rank eight candidates (numbered 1 through 8) in order of preference. The judges submitted their choices in the following manner.

First judge :	5	2	8	1	4	6	3	7
Second judge :z	4	5	7	3	2	8	1	6

(a) Find the coefficient of rank correlation.

(b) Decide how well the judges agreed in their choices.

3. A consumer panel tested nine makes of electric heaters for overall quality. The ranks assigned by the panel and the suggested retail prices were as follows:

Manufacturer	Panel Rating	Suggested Price in Rs.
A	6	480
B	9	395
C	2	575
D	8	550
E	5	510
F	1	545
G	7	400
H	4	465
I	3	420

Is there a significant relationship between the quality and the price of a electric heater?

4. The following data were obtained in a study of relationship between the weight and chest size of infants at birth.

Weight (kg) :	2.75	2.15	4.41	5.52	3.21	4.32	2.31	4.30	3.71
Chest size (cm) :	29.5	26.3	32.2	36.5	27.2	27.7	28.3	30.3	28.7

   (a) Calculate the rank correlation coefficient.

   (b) Test the hypothesis at the 0.025 level of significance that the rank correlation coefficient is zero against the alternative that it is greater than zero.

5. A plant supervisor ranked a sample of eight workers on the number of hours worked overtime and length of employment. Find the rank correlation between two measures. Is it significance at 0.01 level ?

   Amount of overtime: 5.0   8.0   2.0   4.0   3.0   7.0   1.0   6.0

   Years of employment: 1.0   6.0   4.5   2.0   7.0   8.0   4.5   3.0

6. Two judges at a college homecoming parade ranked 8 floats in the following order:

				Float				
	1	2	3	4	5	6	7	8
Judge A	5	8	4	3	6	2	7	1
Judge B	7	5	4	2	8	1	6	3

   (a) Calculate the rank correlation coefficient.

   (b) Test the hypothesis that the rank correlation coefficient is zci" against the alternative hypothesis that it is greater than zero. Use a = 0.05.

Answers

   1. 0.845.

   2. (a) 0.67   (b) The judges did not agree too well in their choices.

   3. r_s = - 0.47; No significant relationship.

   4. (a) r_s= 0.72

   (b) Reject the null hypothesis that the rank correlation coefficient is zero.

   5. r_t = 0.185. The rank correlation is not significant.

   6. (a) r_s= 0.71 *(b)* Reject the null hypothesis

# Median Test for Two Independent Samples

This test is used to test the null hypothesis H_0 that two independent samples have been drawn from identical distribution against the alternative hypothesis that their location parameters (median) are different. The test can be one or two tailed tests. This test is sensitive to differences in location.

Test Statistic

Under the assumption that hypothesis H_0 is true, we can expect roughly 50% of the observations of each sample to be above the median and 50% to be below the median of the combined sample. The sample observations can thus be dichotomized. It is presented in the form of a 2×2 contingency table.

Table 11.3 2 × 2 Contingency Table.

	Sample I	Sample II	Total
Above Median	a	b	a + b
Below Median	c	d	c + d
Total	a + c	b + d	$n = n_1 + n_2$

When $n > 20$ and no cell frequency is less than 5, then

$$X^2 \frac{n\left(lab - bcl - \dfrac{n}{2} \right)^2}{(a+b)(c+d)(b+d)} , \text{ with correction for continuity, is a } X^2 \text{ variable}$$

with I *d.f.*

Example **13 :** An I. Q test was given to a random sample of 15 male and 20 female students of a University. Their scores were recorded as follow:

Male :   56, 66, 62, 81, 75, 73, 83, 68, 48, 70, 60, 77, 86, 44, 72,

Female: 63, 77, 65, 71, 74, 60, 76, 61, 67, 72, 64, 65, 55, 89, 45, 53, 68, 73, 50, 81

Use median test to determine whether I.Q. of male and female students is same in the University. (Given the median of combined sample = 68)

Solution: It is given that the median of the combined sample is 68. On the *discarding two observations with value equal to median, we have n = 33.* The dichotomised observations of the two samples are presented in the following 2 × 2 contingency table :

Table 11.4 2×2 contingency Table.

	Sample I	Sample 11	Total
Above Median	8 (a)	8 (b)	16
Below Median	*6 (c)*	**11 (d)**	17
Total	**14**	**19**	**33**

Thus, we have

$$X^2 \frac{\left[|ab - bc| - (n/2)\right]^2}{(a+b)(c+d)(a+d)(b+d)} = \frac{33(|88 - 48 - 16.5|)^2}{16 \times 17 \times 14 \times 19} = 0.252$$

Tabled value of x^2 for 1 *d.f.* and for $\alpha = 0.05$ is $=3.84$.

Since the tabulated value of $x^2 = 0.252$ is less than the tabled value of $X_{1,0.05} = 3.84$, the null hypothesis is accepted.

Wilcoxon Signed - Rank Test

Wilcoxon Signed - Rank test is useful in comparing two populations (or medians of two populations) for which we have paired observations. Unlike the Sign Test, the Wilcoxon Test accounts for the magnitude of differences between paired values and not only their signs. The test does so by considering the ranks of these differences. The test is, therefore, more efficient than the Sign test when the differences may be quantified rather than just given a positive or negative sign. The Sign Test on the other hand, is easier to carry on.

The Wilcoxon procedure may also be adopted for testing whether the location parameter of a single population (its median or its mean) is equal to any given value. There are one-tailed and two tailed version of each test. We shall discuss the paired observation test for equality of two populations distributions (or the equality of location parameters of the two populations.)

Null hypothesis - H_0: The median difference between the population 1 and 2 is zero.

Alternative hypothesis - H_0: The median difference between the population 1 and2 is not zero

We assume

(i)   that the distribution of differences between two populations is symmetric

(ii)   that the differences are mutually independent.

(iii)   that the measurement scale is at least interval.

By the assumption of symmetry, the hypothesis may be stated in terms of means. The alternative hypothesis may also be directed one : that the mean (or median) of one population is greater than the mean (or median) of the other population.

Steps for Calculation of Test Statistic

(i)   List the pairs of observations we are given for the two populations (or on the two variables)

(ii)   For each pair, calculate the difference: $dc = X. - Y.$

(iii)   Omit all observation *(s)* with equal values and reduce the sample size accordingly

(iv)   Rank these differences in ascending order without regard to their signs.

(v)   The cases of tied ranks are assigned ranks by the average method.

(vi)   Find 2(+) and 2(–), where Σ (+) is the sum of ranks with positive D. and Σ (–) is the sum of ranks with negative D.

(vii)   The Wilcoxon T-statistic is defined as the smaller of the two sums of ranks

$$T = \min (\Sigma(+), \Sigma (-))$$

Where $\Sigma (+) =$ sum of the ranks of positive differences;

$\Sigma (-) =$ sum of the ranks of negative differences

(viii)   *The Decision Rule*: Critical points of the distribution of the test statistic T (when the null hypothesis H_0 is true) are given by the table "Critical values of Wilcoxon T-Test)". We carry out the test on the left tail, i.e. we reject the null hypothesis H_0 if the computed value of the test statistic T is less than the critical point from the table (tabulated value) for a given level of significance.

For One-tailed test, suppose that the alternative hypothesis H, is that mean (or median) of population 1 is greater than that of population 2, *i.e.,*

Null hypothesis $H_0 : \mu_1 = \mu_2$

Alternative hypothesis $H_0 : \mu_1 > \mu_2$

Here we shall use the sum of the ranks of negative differences $\Sigma(-)$

If the alternative hypothesis H_1 is reversed (population 1 and *2* are switched), then we shall use the sum of the ranks of positive differences Σ (+) as the statistic T.

In either case, the test is carried out on the left 'tail' of the distribution. Table "Critical values of Wilcoxon T - test" gives critical points for both one-tailed and two-tailed tests.

Wilcoxon Signed Rank Test - Large Samples Test

In the Wilcoxon test V is defined as the number of pairs of observations from population 1 and 2. As the number of pairs n gets large (as a rule of thumb, $n > 25$ or so), T may the approximate by a normal random variable.

$$\text{Mean of T: E (T)} = \frac{n(n+1)}{4}$$

$$\text{Standard Deviation of T} : \sigma_T = \frac{\sqrt{n(n+1)(2n+1)}}{24} = \text{S.E. (T)}$$

$$\text{Static} : Z = \frac{T - E(T)}{\sigma_T}$$

Tabled Value of Z_α: Find the tabled value of Z_α for a given α-level of significance from table.

Decision: Accept H_0 if calculated value of $|Z|$ < Table value of Z_α otherwise accept the alternative hypothesis H_1

***Example* 14:** Two models of a machine are under consideration for purchase. An organisation has one of each type for trial and each operator, out of the team of 25 operators, uses each machine for a fixed length of time. Their outputs are:

Operator No	: 1	2	3	4	5	6	7	8	9	10	11	12
Output from Machine I	: 82	68	53	75	78	86	64	54	62	70	51	80
Output from Machine II	: 80	71	46	58	60	72	38	60	65	64	38	79
Operator No	: 14	15	16	17	18	19	20	21	22	23	24	25
Output from Machine 1	: 65	70	55	75	64	72	55	70	45	64	58	65
Output from Machine II	: 60	73	48	58	60	76	60	50	30	70	55	60

Is there any significant difference between the output capacities of the two machines? Test at 5% level of significance.

Solution: Let $D = M_1 - M_2$, where M_1 and M_2 denote the outputs of the machines I and II respectively.

Table 11.5 Computation of Ranks.

No.	M1	M2	D	Ranks	No,	M_1	M_2	D_1	Ranks
1	82	80	2	2	14	65	60	5	10
2	68	71	-3	4.5	15	70	73	-3	4.5
3	53	46	7	15.5	16	55	48	7	15.5
4	75	58	17	20.5	17	75	58	17	20.5
5	78	60	18	22	18	64	60	4	7.5
6	86	72	14	18	19	72	76	-4	7.5
7	64	38	26	24	20	55	60	-5	10
8	54	60	-6	13	21	70	50	20	23
9	62	65	-3	4.5	22	45	30	15	19
10	70	64	6	13	23	64	70	-6	13
11	51	38	13	17	24	58	55	3	4.5
12	80	79	1	1	25	65	60	5	10
13	64	37	27	25					

$$\Sigma(+) = 2 + 15.5 + 20.5 + 22 + 18 + 24 + 13 + 17 + 1 + 25 + 10 + 15.5 + 20.5 + 7.5 + 23 + 19 + 4.5 + 10 = 268$$

$$\Sigma(-) = 4.5 + 13 + 4.5 + 4.5 + 7.5 + 10 + 13 = 57$$

Null hypothesis H_0 : Output capacities of the two machines are same.

Alternative hypothesis H_1 : Output capacities are not same.

T - Statistic= min $(\Sigma(+), \Sigma(-)) = 57$

$$\therefore \quad \mu_T = \frac{n(n+1)}{4} - \frac{25 \times 26}{4} = 162.5$$

$$\alpha_T = \sqrt{\frac{n(n+1)(2n+1)}{24}} = \sqrt{\frac{25 \times 26 \times 51}{24}} = 37.2$$

$$\therefore \quad \text{Test Statistics} : Z = \frac{T - E(T)}{\sigma_T} = \frac{57 - 162.5}{24} = 37.2$$

$\therefore$ Critical Value: Tabled value of Z_u for $\alpha = 5\% = 1.96$.

Since $|Z| > Z_\alpha$ $(\alpha = 5\%$ level), so H_0 is rejected at 5% level of significance.

Hence the output capacities are not same at 5% level of significance.

***Example* 15:** A random sample of 30 students obtained the following marks in a class test. Test the hypothesis that their median score is more than 50.

55 58   25   32   26   85   44   80   33   72   10   42   15   46   64

39 38   30   36   65   72   46   54   36   89   94   25   74   66   29

Solution:

Null hypothesis H_0 : $M_d = 50$

Alternative hypothesis H_1: $M_d > 50$

Let $D = x - 50$.

The D values and their rankings are shown in the following table :

D	:	5	8	-25	-18	-24	35	-6	30	-17	22
Rank	:	4	6.5	23.5	16	21.5	26.5	5	25	15	19.5
D	:	-40	-8	-35	-4	14	-11	-12	-20	-14	15
Rank	:	29	6.5	26.5	2	11	8	9	17	11	13
D	:	22	-4	4	-14	39	44	-25	24	16	-21
Rank	:	19.5	2	2	11	28	30	23.5	21.5	14	18

Here $n = 30$.

$\Sigma(+)$ = Sum of ranks with positive D = 220.5

$\Sigma(-)$ = Sum of ranks with negative D = 244.5

Let us take T-Statistics as T = $\Sigma(-)$ = 244.5.

[$\because$ It is a one tailed test so T is taken as the sum of ranks with negative D, i.e., T = (–)]

$$\therefore \quad \text{Mean } \mu_T = \frac{n(n+1)}{4} = \frac{30 \times 31}{4} = 232.5$$

Standard Error :

$$\sigma_T = \sqrt{\frac{n(n+1)(2n+1)}{24}} = \sqrt{\frac{30 \times 31 \times 16}{24}} = \sqrt{2363.75} = 48.62$$

Test Statistics : $Z = \dfrac{T - \mu_T}{S.E(T)} = \dfrac{244 - 232.5}{48.62} = 48.62 = 0.247$

Tabled value of Z_α at $\alpha < x = 5\%$ level of significance = 1.645.

Result: Since $0.2477 < 1.645 \Rightarrow Z < Z_\alpha \Rightarrow$ the null hypothesis H_0 is accepted at 5% level of significance. The median score is 50.

Chapter 12

Correlation, Regression and Probit Analysis

Correlation and regression analysis are two statistical methods employed to determine if their exists any relationship between two variables and to express their relationship mathematically or numerically. If for every measurement of a variable x we know a corresponding value of a second variable y, the resulting set of pairs of variables is called bivariate population.

Examples of bivariate population:

1. Heights (x) and weights (y) of students in a class.

2. Packed cell volume (x) and red blood cell count (y).

3. Concentration of a chemical substance (x) and absorbance or optical density (y)

4. Temperature (x) and solubility (y)

5. Temperature (x) and rate of reaction (y)

6. Time (x) and amount remained in drug decomposition (y)

Two variables (x and y) are said to be correlative if an increase in one variable is accompanied by an increase or decrease of the other. If higher values of one variable are associated with higher

LEARNING OBJECTIVES

♦ To know

♦ Correlation and correlation coefficient

♦ Estimation of correlation coefficient with worked out examples.

♦ Regression analysis and method of least square

♦ Worked out examples.

♦ Multiple regression

♦ Probit analysis, determination of LD $_{50}$ and ED $_{50}$

values of the other or when lower values of one are accompanied by the lower values of the other (i.e when the movement of the two variables are in one direction) they are said to have positive or direct correlation.

Ex: The greater the radius of the circle, the greater will be its circumference.

If the higher values of one variable are associated with the lower values of the other (i.e when the movements of the two variables are in the opposite direction) , the variables are said to have negative or inverse correlation.

Examples:

1. Time and amount remained in drug decomposition.

2. Temperature and reaction time.

3. In the case of gases, volume varies inversely with the pressure at constant temperature

The strength or degree of the relationship between two variables x and y is measured by Karl Pearson correlation coefficient (r).

The correlation coefficient (r) for a sample of n pairs of x, y values is defined as

$$r = \Sigma xy / \sqrt{\Sigma x^2 \Sigma y^2}$$

For the purpose of computation, it is given in the following alternate form

$$r = [xy - (x)(y)/n] / \Sigma x^2 - \frac{(\Sigma x)^2}{n(\Sigma y)} 2 - (y)^2 / n$$

(or)

$$r = xy - n\bar{x}y$$

Properties and Interpretation of 'r'

The value of 'r'is independent of the origin of reference and the unit of measurement. It can take any value from -1 to +1 depending on the nature of the relationship of the given variable. A value of $r = \pm1$ implies that the variables are perfectly related. The magnitude of the correlation coefficient 'r', determines the strength of the relationship. Whereas the sign indicates whether the variables are positively correlated or negatively correlated.

$r = 0$ indicates no correlation but does not necessarily imply that the variables are independent.

Scatter Diagram

A plot of dependent variable (y) against the independent variable (x) is known as Scatter diagram. The pattern or the way in which the plotted points lie on the scatter diagram reveals the type of correlation. If the correlation was perfect all the points lie on the straight line indicating a definite relationship.

Causation and Correlation

The presence of correlation between two variables does not necessarily imply the existence of direct causation, though causation will always result in correlation. The observed correlation, may be due to any one of the following:

1. One variable being the cause of the other. That variable which is the cause is called "subject" or independent variable and is usually taken on x-axis and the one which is the effect is called relative or dependent variable and is taken on the y-axis.

    Example: Concentration (independent variable) Vs Absorbance (dependent variable)

2. Both variables being the result of a common cause or a third factor.

    Example: The positive correlation between yield per acre of rice and of jute is due to the fact that the two are related to the amount of the rain fall.

3. *Chance*: A fair degree of correlation may be observed between two variables in a sample when it does not exist when in the whole of the population by chance. Such as correlation is known as spurious correlation.

Worked-out Examples

Example 1: Calculate the correlation between x and y based on the data given below

x	Y	xy	x^2	y^2
12	14	168	144	196
9	8	72	81	64
8	6	48	64	36
10	9	90	100	81
11	11	121	121	121
13	12	156	169	144
7	3	21	49	9
$\Sigma=$		676	728	651

$$r = \sum xy / \sqrt{\sum x^2 \sum y^2}$$

$$r = 676 / \sqrt{(728 \times 651)}$$

$$= 676/688.42 = 0.981$$

Hence the correlation between x and y is 0.981.

***Example* 2:** Find the correlation between x and y from the following data

	X	**Y**	x^2	**xy**	y^2
	10	3	100	30	9
	20	2	400	40	4
	30	0	900	0	0
	40	5	1600	200	25
	50	4	2500	200	16
Σ	150	14	5500	470	54

$$r = \sum xy / \sqrt{\sum x^2 \sum y^2}$$

$$r = 470 / \sqrt{5500 \times 54}$$

$$= 470/544.97$$

$$= 0.86$$

Hence the correlation between x and y is 0.86.

Coefficient of Rank Correlation

Correlation coefficient (r) cannot be used in cases where the direct quantitative measurement of the phenomenon under study is not possible.

Examples: efficiency, honesty, intelligence etc..,

In such cases one may rank the different items and apply the Spearman's method of rank differences for finding out the degree of correlation.

$$R = 1 - 6\sum D^2 / N (N^2 - 1)$$

Where D = difference between paired ranks

N = no. of pairs

When correlation between attributes or characteristics of population is needed the Spearman's rank correlation is used.

***Example* 3:** The following are the data obtained on DT and hardness of various tablet formulations. Find the correlation between DT and Hardness.

In this example we have to estimate the correlation between two attributes namely DT and Hardness of the tablets. Rank correlation coefficient (R) can be estimated based on the data as follows.

Ranks are assigned separately to the observations of DT and Hardness, Fast disintegrating tablet is given rank 1 and tablet with low hardness is given rank 1. The analysis is shown in the following table.

Tablet	DT (min)	Rank assigned to DT	Hardness (kg/sq.cm)	Rank assigned to Hardness	Difference in ranks (D)	D^2
1	0-50	1	5.0	1	0	0
2	1-30	3	7.5	3	0	0
3	18-20	6	9.0	5	1	1
4	18-00	5	13.0	6	-1	1
5	0-55	2	7.0	2	0	0
6	17-18	4	8.0	4	0	0 Total = 2

$$R = 1 - 6 \, \Sigma D^2 / N \, (N^2 - 1)$$

$$= 1 - 6 \, (2) / 6 \, (6^2 - 1)$$

$$1 - 12/ 6 \, (35)$$

$$1 - 12/210$$

$$1 - 0.057$$

$$= 0.943$$

Hence the correlation coefficient (R) between DT and Hardness of the tablets based on the given data was found to be 0.943

Regression Analysis

Regression is the process in which relationship between two variables in terms of an equation is established so that the value of the variables can be provided for a given value of the other. The problem of linear regression involves fitting a straight line to the points plotted in a Scatter diagram.

Let y denote the dependent variable and x the independent variable. In linear regression the relationship is linear and it will be of the form y = a+ bx. The straight line will not pass through all the points of the scatter diagram. The spread of the plotted points about the straight line gives rise to errors in the prediction of the value of the dependent variable from that of the independent variable. Reliable prediction requires that the dispersion about the straight line shall be small.

The best fit line to a scattered diagram of 'n' points is obtained by the method of least squares.

Method of Least Squares

The method of least squares is a method of fitting a straight line to a set of 'n' points in a such a way that $\Sigma (y - y_e)^2$ has its smallest possible value, where the sum is calculated for the given 'n' pairs of values of x and y. 'y' is the actual value and 'y_e' is the predicted value on the regression line.

The least squares method of fitting a straight line $y = a + bx$ to the given data of x and y consists of choosing the values of 'a' and 'b' such that the sum of the squared deviations, $\Sigma (y - y_e)^2 = \Sigma (y - a - bx)^2$ is the least. The constants 'a' and 'b' under this method are obtained by solving the following two normal equations.

$$\Sigma y = na + b \Sigma x$$

$$\Sigma xy = a \Sigma x + b \Sigma x^2$$

These equations are solved for 'a' and 'b'

***Example* 4:** Find the regression equation of y on x from the following data

x	Y	xy	x^2	y^2
12	14	168	144	196
9	8	72	81	64
8	6	48	64	36
10	9	90	100	81
11	11	121	121	121
13	12	156	169	144
7	3	21	49	9
$\Sigma = 70$	63	676	728	651

For developing regression equation we have to solve to following two normal equations to obtained the constant 'a' and 'b'

$$\Sigma y = na + b \Sigma x$$

$$\Sigma xy = a \Sigma x + b \Sigma x^2$$

Based on the given data the two above equations are as follows.

$$63 = 7a + 70b \qquad .....(i)$$

$$676 = 70a + 728b \qquad .....(ii)$$

By multiplying equation (i) with 10 we get,

$$630 = 70a + 700b \qquad .....(iii)$$

Equation (ii) minus equation (iii) =

$$46 = 28b$$

Hence , b = 46/28 =1.6

Substitute 'b 'value in equation (i) and solve for 'a'

$$63 = 7a + 70 \times 1.6$$

$$7a = -52$$

$$a = -52/7 = -7.42$$

Hence the regression equation is **y = 1.6 × −7.42**

 ## Multiple Regression

Linear Regression between Three Variables

The relationship between 1 dependent variable y and two independent variable x_1 and x_2 can be expressed by the linear equation of the type

$$y_i = \beta_0 + \beta_1 x_{1i} + \beta_2 x_{2i}$$

Multiple regression is the analysis in which a mathematical relationship is established between three variables, two independent variables x_1 and x_2 and one dependent variable y. The dependent variable 'y' is influenced by two independent variables x_1 and x_2. In such cases the mathematical regression equation will be of the type

$$y_i = \beta_0 + \beta_1 x_{1i} + \beta_2 x_{2i}$$

Where x_1 and x_2 are two independent variables and y is the dependent variable which depends on both x_1 and x_2.

β_1 and β_2 are the partial regression coefficients and β_0 is the intercept term.

β_1 measures the change in y for a unit change in x_1 when x_2 is kept constant. Similarly β_2 measures the change in y for unit change in x_2 when x_1 is kept constant. The constants can be estimated with the help of the principle of the least squares i.e. minimizing the residual sum of squares i.e $\sum(Y_i - Y_e)^2$ is minimum.

Quantities of β_0, β_1 and β_2 can be estimated by solving the following three normal equations

$$\sum Y_i = n\,\beta_0 + \beta_1 \sum X_{1i} + \beta_2 \sum X_{2i}$$

$$\sum X_{1i} Y_i = \beta_0 \sum X_{1i} + \beta_1 \sum X^2_{1i} + \beta_2 \sum X_{1i} X_{2i}$$

$$\sum X_{2i} Y_i = \beta_0 \sum X_{2i} + \beta_1 \sum X_{1i} X_{2i} + \beta_2 \sum X^2_{2i}$$

On solving the above normal equations we can arrive at the following expressions for β_0, β_1 and β_2.

$$\beta_0 = y - \beta_1 \bar{x}_1 - \beta_2 \bar{x}_2$$

$$\beta_1 = [(\Sigma x_1 y)(\Sigma x_2^2) - (\Sigma x_2 y)(\Sigma x_1 x_2)] / (\Sigma x_1^2)(\Sigma x_2^2) - (\Sigma x_1 x_2)^2$$

$$\beta_2 = [(\Sigma x_2 y)(\Sigma x_1^2) - (\Sigma x_1 y)(\Sigma x_1 x_2)] / (\Sigma x_1^2)(\Sigma x_2^2) - (\Sigma x_1 x_2)^2$$

Examples of Multiple Regression

1. Yield of paddy (Y) is dependent on two independent variables such as quantity of the fertilizer applied (X_1) and amount of insecticide used (X_2)

2. Yield of a chemical reaction (Y) depends on two independent variables such as temperature (X_1) and pH (X_2).

3. The dissolution rate of tablet (Y) depends on two independent variables such as concentration of binder (X_1) and concentration of disintegrant (X_2).

4. Yield of an antibiotic in an aerobic fermentation process (Y)depends on two independent variables such as pH (X_1) and Aeration rate (X_2).

In all the above cases, the dependent variable (Y) depends on two independent variables X_1 and X_2.

In the experiments to evaluate the effect of two factors (independent variables) on the dependent variable , combination of different levels of independent variables were used and in each case the dependent variable is measured giving rise a set of data relating the three variables. In such cases multiple regression is needed to establish the relationship between the dependent variable (Y) and independent variables X_1 and X_2.

 **Probit Analysis and its Applications**

Probit analysis is a statistical technique used to convert the sigmoid or S- shaped dose-response curve to a straight line that can be used to compare the potencies and to determine lethal dose (LD_{50}) and effective dose (ED_{50}) of therapeutic substances. It has the application in bioassays and screening methods in which the responses are of all-or-none type, i.e death or no death, effect or no effect etc.

The relationship between the response and doses, in such cases are of sigmoid or S –shaped curves as shown below.

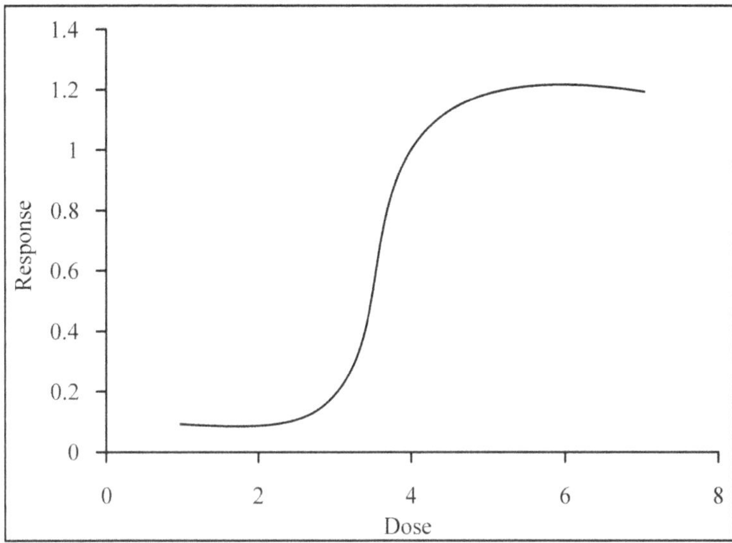

Figure 12.1

A Typical Dose-Response Curve

A form of linearization is required for making evaluations in such cases. A widely accepted linearization is to convert the dose into log dose and response into probits.

Probit is an abbreviation for probability. Probits are derived from the standard deviation units in the table of the standard normal probability curve. The 50 % response is equivalent to the left hand half of the area under the normal curve upto standard deviation unit of zero. Since the probits are obtained by adding 5 to standard deviation unit 'Z', a 50% response corresponds to a probit of 5.

A probit is defined as follows

Probit $=$ Z+5 where Z is the standard normal variate or standard deviation unit corresponding to the percentage response.

Probits corresponding to various percentage responses are available in statistical tables. For example, a 16% percent response corresponds to a probit of 4.006. A 97.5% response corresponds to a standard deviation unit (Z) of 2 and hence a probit of 2+5=7

Determination of LD$_{50}$

LD$_{50}$ is the lethal dose to the 50 percent of the animals. For determining LD$_{50}$, the response measured is death or no death after administration of increasing doses to group of animals like mice or rats. Initially the least tolerated (smallest dose) and most tolerated (highest dose) are found out by hit and trail method. Five doses are then selected in this range of doses and administered to groups of animals each containing atleast 10 animals after overnight fasting. The drug is normally injected by intra-peritoneal route or by oral route and the animals are observed for 2 hours for death due to acute toxicity. Number of deaths are noted in each group and then % mortality or deaths are calculated.

The % deaths i.e response is converted into probits using probit-statistical tables. Dose is converted into log dose. A plot of log dose on x-axis and response in probits on y-axis is drawn. The plot is normally a straight line as follows:

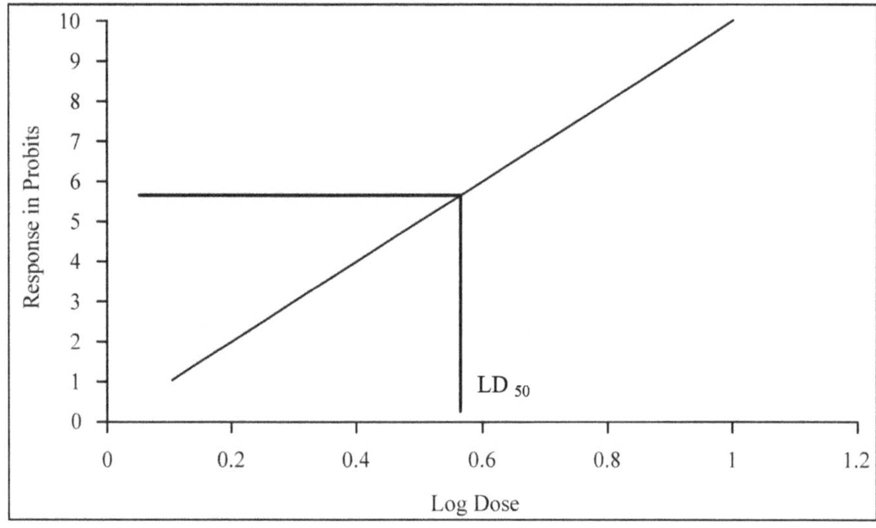

Figure 12.2

From the linear plot, dose corresponding to a probit value of 5 is recorded. This is the dose lethal to 50% animals i.e LD$_{50}$.

Determination of ED$_{50}$

ED$_{50}$ is the effective dose to 50 % percent animals. For determining the ED$_{50}$, the response or the effect should be clearly defined initially. For example, when an anti-diabetic drug or a herbal extract is evaluated for its anti- diabetic activity, a reduction of 25% percent in blood sugar levels is taken as a measure of activity or effect. This response i.e 25% reduction in blood glucose is measured after administering increasing doses to groups of animals such as rats or mice. When the blood glucose reduction is less than 25 %, it is recorded as 'no effect'. When the glucose reduction is more than 25% it is recorded as 'effect'.

The response is thus measured as effect or no effect. Increasing doses of the drug are administered to groups of animals each consisting atleast 10 animals after overnight fasting. The drug is administered by either intraperitoneally or by oral route. The animals are then observed for 4-12 hours for the effect or no effect. The numbers of animals showing the effect are noted in each group and then the percentage response in each group is calculated. Percentage response is converted into probits using statistical probit tables. The dose is converted into log dose. A plot of log dose on x-axis and probits corresponding to percentage (%) response on y-axis is drawn. The plot is normally a straight line as follows.

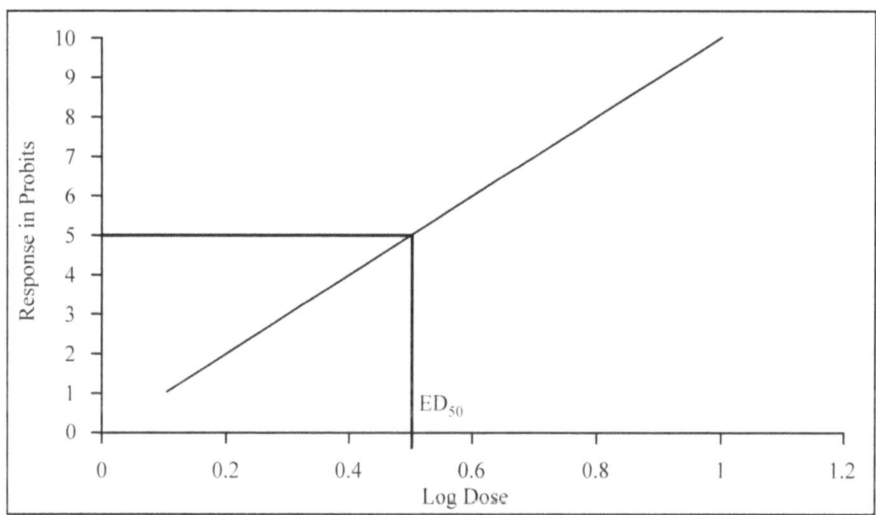

Figure 12.3

From the linear plot the dose corresponding to a probit value of 5 is recorded. This is the dose effective to 50 percent animals i.e. ED$_{50}$

Chapter 13

Research Methodologies - VI
(Patentable Research in Pharmacy)

IPR and Patents

Intellectual property rights (IPRs) are the rights given by the law of a country to persons over the creations of their minds. They usually give the creator an exclusive right over the use of his/her creations for a certain period of time. The following are the different types of IPRs.

Patents

Patent is a monopoly right granted by the Government to exclude others from exploiting or using a particular invention

Copyrights

Copyright is a legal right that grants the creator of an original work exclusive right to use and distribution for a limited time with the intention of enabling the creators to receive compensation for their intellectual effort

Trademarks

A trademark generally refers to a brand or logo. Trademarks can also be obtained for a business name, distinctive catch phrases, captions etc.

LEARNING OBJECTIVES

- To know
- Patents, types of patents and procedure of obtaining patents.
- Requirements of patentable research
- Study of patents related to pharmacy with six examples.

Industrial Designs

A design registration is used to obtain protection for the features of shape, configuration, pattern or ornaments.

Patents and Patentable Research

Patent is an exclusive monopoly granted by the government to an inventor over his invention for a limited period of time. Patents are issued for any invention, whether product or process, in all fields of technology, provided that they are New, involve an inventive step and are capable of industrial application. A patent describes an invention and creates a legal right under which the patented invention can normally be exploited (i.e., manufactured, used, sold, imported) by the owner of the patent. The right created by a patent is a monopoly right. The significance of the right is a statutory right to prevent others from exploiting his invention.

Object of Patent Grant System

The object of granting a patent is the encouragement and development of new technology and industry in the country. Patents are given only for inventions. Inventions are solutions to specific problems in the field of technology. An invention may relate to a product or a process. Invention is a property created with the use of intellect

Patentability: In order to get a patent for an invention, the invention has to be patentable.

There are some exceptions to patentability. The following are not patentable

(i)   Discoveries of materials already in nature are not patentable. These are known to all and fall within public domain

(ii) Scientific theories or mathematical models are not patentable. These theories are not inventions. However inventions made using these theories are patentable.

(iii) Biological processes are not patentable. In order to qualify for patent protection, the substance has to be manmade.

(iv) Methods of treatments and diagnostic methods for human/ animals are not patentable. These are very subjective and cannot be standardized by patents.

Requirements of Patentability

- Novelty
- Non-obviousness / inventive step
- Industrial applicability.

Novelty

Novelty is a fundamental requirement for grant of patent. Novelty means that the invention should be new. What is not new is said to exist already in the public domain, therefore not patentable. What is already in public domain is called "prior art". Prior art is all the knowledge that existed prior to the relevant filing date of a patent application. Novelty exists if there is any difference between the invention and the prior art.

Non-obviousness

It means that the invention shall possess sufficient 'inventive step' in such a manner that the invention is non-obvious to one skilled in the same art. Inventive step is the result of a creative idea. It must be a noticeable step. There must be a clearly noticeable difference between the state of the art and the claimed invention. There must be an advance or progress over the prior art.

Industrial Applicability

The subject matter should have some practical application and it can be reproduced for industrial application.

Types of Patents

(i)   *Process Patent*: Nobody has right to prepare products in that particular process except patentee. (1970----31/12/2004).

(ii)   *Product Patent*: Nobody has the right to prepare that particular product except the patentee. (1/1/2005---till date).

Impact of Process Patents

It allows reverse engineering of drugs. It gives more flexibility for production of drugs. This is more cost effective regime. It suppresses creativity for newer and better products.

Impact of Product Patents

It does not allow reverse engineering of drugs. It allows a patent span of 20 years. It gives a chance for young and dynamic individuals for a better chance for application of their intellects for newer and better products. It increases the prices of drugs alarmingly since there is no competitor.

Changes Observed after Product Patent

Research results are patented before publication. Corporate and laboratories collaborated on R & D basis. Research become commercialized. Technologies developed are licensed out.

 # Procedure for Obtaining Patent

The following are stages for obtaining a patent

  (i)   Application

  (ii)   Provisional specification

  (iii) Complete specification

  (iv) Examination report

  (v)   Time for compliance

  (vi) Gazette publication

  (vii) Objections

  (viii) Patent certificate issue

Application

An application for a patent may be made by the inventor, his legal representatives, or his assignee, either alone or jointly with any other person. An application for a patent of addition may be made only by the applicant for original patent to which it is an addition. Every application for a patent shall be for one invention only and shall be made in the prescribed form and filed in the Patent Office (Head Office at Calcutta and Branch Offices at Madras, Bombay and Delhi).

Provisional Specification

Every application must state that the applicant is in possession of the invention and must have the owner claiming to be true and first inventor, and where the person so claiming is not the applicant, the application believes the person so named to be true and first inventor. A provisional specification is a document in a prescribed form containing a description of the essential features of the invention.

Complete Specification

A complete specification is a document drawn in a prescribed form and contains the following:

  (i)   A full description of the invention and its operation or use and the method of performance.

  (ii) Disclosure of the best method of performing the invention known to the applicant for which he is entitled to claim.

  (iii) A statement of claim(s) defining the scope of invention for which protection is sought.

  (iv) Where an application is accompanied by only a provisional specification, a complete specification must be filed within 12 months from the date of filing of the application.

Examination Report

On receiving the application for a patent along with complete specification, the Controller shall refer it to an examiner for making a report to him in respect of whether the application and specifications are in accordance with the requirements of the Act; Whether there is any lawful ground of objection to the grant of patent under the Act in pursuance of application.

Time for Compliance

Examiner's report should be obtained within 18 months. If the report is adverse to the applicant or requires any amendment, the Controller shall communicate the list of the objections to the applicant and give him an opportunity of being heard. The Controller is also empowered to refuse or require amend application in certain cases.

Gazette Publication

On the acceptance of a complete specification, the Controller shall notify the applicant and advertise in the Official Gazette the fact that the specification has been accepted and then the specifications shall be open to public inspection.

Objections

Any person may oppose the grant of patent within a period of 4 months from the date of advertisement of acceptance, on any of the following grounds;

(i)   That the invention or part thereof has been obtained by the applicant wrongfully,

(ii) That the invention in question is obvious and clearly does not involve any inventive step;

(iii) That the subject of any claim of the complete specification is not an invention within the meaning of this Act or is not patentable under this Act;

(iv) That the complete specification does not sufficiently and clearly describe the invention or the method by which it is performed

Patent Certificate Issue

On receiving any such notice of opposition, the Controller shall notify the applicant and give him and the opponent an opportunity to be heard before deciding the case. The Controller is empowered to refuse or grant the patent. Once the patent, is granted, the Controller shall cause it to be sealed with the seal of the patent office and the date of sealing shall be entered in the register.

Examples of Patentable Research in Pharmacy

Example 1

Title: Oral Formulations of Cladribine

- Invention relates to a composition comprising a complex, cladribine-cyclodextrin complex formulated into a solid oral dosage form and to a method for enhancing the oral bioavailability of cladribine

- Date of application : 6-10-2005

- Date sanction : 29-06-2012

- Cladribine: Immunosuppressive agent

  An acid labile drug

  Very low oral bioavailability (20-25%)

- An earlier patent describes complexation and solubilizing of cladribine with cyclodextrins for formulation of i.m/s.e injections

- Another report : Cladribine is more stable against acid catalyzed hydrolysis when combined with cyclodextrin

- No reports on solid dosage forms and enhancement of oral BA

- The patent describes detailed procedures of

- Phase solubility study with HP β CD

- Preparation of cladribine - HP β CD complex for human use

- Method used : Lyophilization to produce amorphous form

- Weight ratios of 1:10 to 1:16

- Characterization by HPLC, MS,DSC, XRD

- Preparation of oral tablets : Manufacturing details of two different strengths and their quality tests

- Clinical studies : To establish dose response, absolute BA and relative BA

- Three oral formulations in different doses, one i.v injection and one s.c injection; n = 38 patients with multiple sclerosis

- Plasma analysis by LC-MS

- Pharmacokinetic data analysis by statistical soft ware

  Claims:

- A complex cladribine-cyclodextrin complex formulated into a solid oral dosage form

- Cladribine-cyclodextrin complex characterized in that it consists of an intimate amorphous admixture of (a) cyclodextrin and (b) amorphous cladribine

- The said dosage form comprises no significant amount of free crystalline cladribine therein

- Said dosage form having a weight ratio of cladribine to said amorphous cyclodextrin of from 1:10 to 1:16.

  A process for the preparation of a complex cladribine- cyclodextrin complex which is characterized by the steps of :

- Combining cladribine and the amorphous cyclodextrin, hydroxypropyl-β- cyclodextrin in water at a temperature of from 40° C, to 80° C and maintained said temperature for a period of from 6hrs to 24hrs, wherein said cladribine and said cyclodextrin are present in sufficient quantities to produce an amorphous product having a weight ratio cladribine to amorphous cyclodextrin of from 1:10 to 1:16.

- Cooling the resultant aqueous solution to room temperature

- Lyophilizing the cooled solution to afford an amorphous product

  A process for preparation of a complex cladribine-cyclodextrin complex formulated into a solid oral dosage form which is characterized by steps of

- Combining cladribine and the amorphous cyclodextrin hydroxypropyl-β-cyclodextrin in water and lyophilization

- Formulating the amorphous product into a solid oral dosage form

- The process comprises blending the complex with magnesium stearate and compressing into tablet

- Magnesium stearate is pre mixed with sorbitol powder before blending with the complex

***Example* 2:** Starch – Urea – Borate Matrix Compositions for Controlled Release of Active Pharmaceutical Ingredients

- The investigation relates to the field of Pharmaceutical Technology and describes preparation of starch – urea – borate, a modified cross-linked starch and novel matrix compositions employing starch-urea-borate for obtaining controlled release of active pharmaceutical ingredients (APIs).

- The novel starch-urea-borate matrix composition is suitable for producing controlled release preparations of a large number of APIs.

- In the new matrix composition as active ingredient is present essentially uniformly dispersed in a matrix composed of starch-urea-borate with or without other additives such as diluents and solubilizers.

- The matrix compositions are in the form of compressed tablets as a platform technology for obtaining controlled release of APIs belonging to different chemical and pharmacological categories and which require controlled release formulation.

- Matrix compositions employing starch-urea-borate are suitable for once a day and twice a day controlled release of several APIs.

***Example* 3:** Olibanum Resin as Microencapsulating Agent for Controlled Release of Active Pharmaceutical Ingredients

- The investigation relates to the field of Pharmaceutical Technology and describes olibanum resin, a natural lipophilic polymer as microencapsulating agent and microencapsulation of active pharmaceutical ingredients by the olibanum resin for obtaining controlled release.

- The invention also describes novel controlled release microcapsule compositions employing olibanum resin for obtaining controlled release of several APIs.

- The novel controlled release microcapsule compositions are presented as spherical, discrete, free-flowing in size range 300-800µm, in which the API is present essentially uniformly dispersed in olibanum resin polymer with or without other additives such as diluents and solubilizers.

- The novel microcapsule compositions employing olibanum resin are suitable for producing controlled release of a large number of APIs belonging to different chemical and pharmacological categories as platform technology for obtaining controlled release of APIs.

- The microcapsule compositions developed in the invention are suitable for once a day and twice a day controlled release of several APIs.

***Example* 4:** Starch Acetate as Microencapsulating Agent for Controlled Release of Active Pharmaceutical Ingredients

- The investigation relates to the field of Pharmaceutical Technology and describes the preparation and application of starch acetate, a chemically modified lipophilic starch polymer as microencapsulating agent and microencapsulation of active pharmaceutical ingredients by starch acetate for obtaining controlled release.

- The invention also describes novel controlled release microcapsule compositions employing starch acetate for obtaining controlled release of several APIs.

- The novel controlled release microcapsule compositions are presented as spherical, discrete, free-flowing in size range 300-800μm, in which the API is present essentially uniformly dispersed in starch acetate polymer with or without other additives such as diluents and solubilizers.

- The novel microcapsule compositions employing starch acetate are suitable for producing controlled release of a large number of APIs belonging to different chemical and pharmacological categories as platform technology for obtaining controlled release of APIs.

- The microcapsule compositions developed in the invention are suitable for once a day and twice a day controlled release of several APIs.

***Example* 5:** Calcium Starch Based Matrix Compositions for Controlled Release of Active Pharmaceutical Ingredients

- The investigation relates to the field of Pharmaceutical Technology and describes preparation of calcium starch, a modified cross linked starch and novel matrix compositions employing calcium starch for obtaining controlled release of active pharmaceutical ingredients (APIs).

- The novel calcium starch matrix composition is suitable for producing controlled release preparations of a large number of APIs.

- In the new matrix composition an active ingredient is present essentially uniformly dispersed in matrix composed of calcium starch with or without other additives such as diluents and solubilizers.

- The matrix compositions are in the form of compressed tablets as a platform technology for obtaining controlled release of APIs belonging to different chemical and pharmacological categories and which require controlled release formulation.

- Matrix compositions employing calcium starch are suitable for once a day and twice a day controlled release of several APIs.

***Example* 6:**

Title: Pharmaceutical Composition

Pharmaceutical composition which comprises an insulin sensitivity enhancer in combination with other antidiabetics differing from the enhancer in the mechanism of action, which shows a potent depressive effect on diabetic hyperglycemia and is useful for prophylaxis and treatment of diabetes.

Table 13.1 Pharmaceutical Composition.

1.	Piogltiazone hydrochloride	10 mg
2.	Glibenclamide	1.25 mg
3.	Lactose	86.25 mg
4.	Corn starch	20 mg
5.	Polyethylene glycol	2.5 mg
6.	Hydroxypropylcellulose	4 mg
7.	Carmellose calcium	5.5 mg
8.	Magnesium stearate	0.5 mg
		130 mg (per tablet)

The whole amounts of (1), (2), (3), (4) and (5), 2/3 amounts of (6) and (7), and ½ amount of (8) are mixed well and granulated in the conventional manner. Then, the balances of (6), (7) and (8) are added to the granules, which is mixed well and the whole composition is compressed with a table machine. The adult dosage is 3 tablets/day, to be taken in 1 to 3 divided doses.

 **Conclusion**

Patentable research in pharmacy is majorly carried out in pharma industries and research organizations. Patentable research in pharmacy institutions at all levels is negligible, which needs encouragement and support. Patent cells/units may be established in all University level institutions to cater to the needs of patentable research.